T0211530

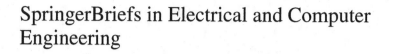

SpringerBriefs in Electrical and Computer Engineering

More information about this series at http://www.springer.com/series/10059

Guodong Zhao • Wei Zhang • Shaoqian Li

Advanced Sensing Techniques for Cognitive Radio

Springer

Guodong Zhao
University of Electronic Science
    and Technology of China
Chengdu, China

Wei Zhang
The University of New South Wales
Sydney, NSW, Australia

Shaoqian Li
University of Electronic Science
    and Technology of China
Chengdu, China

Material for chapters 2–4 originated from previous works:
Chapter 2 – L. Li, G. Zhao, and X. Zhou, "Enhancing small cell transmission opportunity through passive receiver detection in two-tier heterogeneous networks," IEEE Trans. Signal Process., vol. 63, no. 13, pp. 3461-3473, Jul. 2015. Reprinted with permission from IEEE.
Chapter 3 – G. Zhao, G. Li, and C. Yang, "Proactive detection of spectrum opportunities in primary systems with power control," IEEE Trans. Wireless Commun., vol. 8, no. 9, pp. 4815–4823, Sept. 2009. Reprinted with permission from IEEE.
Chapter 4 – G. Zhao, B. Huang, L. Li, and X. Zhou, "Relay-assisted cross-channel gain estimation for spectrum sharing," IEEE Trans. Commun., vol. 64, no. 3, pp. 973–986, Mar. 2016. Reprinted with permission from IEEE.

ISSN 2191-8112   ISSN 2191-8120   (electronic)
SpringerBriefs in Electrical and Computer Engineering
ISBN 978-3-319-42783-6   ISBN 978-3-319-42784-3   (eBook)
DOI 10.1007/978-3-319-42784-3

Library of Congress Control Number: 2016948292

© The Author(s) 2017
This work is subject to copyright. All rights are reserved by the Publisher, whether the whole or part of the material is concerned, specifically the rights of translation, reprinting, reuse of illustrations, recitation, broadcasting, reproduction on microfilms or in any other physical way, and transmission or information storage and retrieval, electronic adaptation, computer software, or by similar or dissimilar methodology now known or hereafter developed.
The use of general descriptive names, registered names, trademarks, service marks, etc. in this publication does not imply, even in the absence of a specific statement, that such names are exempt from the relevant protective laws and regulations and therefore free for general use.
The publisher, the authors and the editors are safe to assume that the advice and information in this book are believed to be true and accurate at the date of publication. Neither the publisher nor the authors or the editors give a warranty, express or implied, with respect to the material contained herein or for any errors or omissions that may have been made.

Printed on acid-free paper

This Springer imprint is published by Springer Nature
The registered company is Springer International Publishing AG Switzerland

Preface

With the exponential growth of wireless data services, the spectrum shortage becomes extremely severe. As a promising and new technology to break the spectrum gridlock, cognitive radio has received much attention in both academia and industry. In cognitive radio, spectrum sensing is crucial since it identifies the spectrum holes for secondary user transmission. This Springer Brief investigates advanced sensing techniques to detect and estimate the primary receiver for cognitive radio systems. Along with a comprehensive overview of existing spectrum sensing techniques, this Brief focuses on the design of new signal processing techniques, including the region-based sensing, jamming-based probing, and relay-based probing. The proposed sensing techniques aim to detect the nearby primary receiver and estimate the cross-channel gain between the cognitive transmitter and primary receiver. The performance of the proposed algorithms is evaluated by simulations in terms of several performance parameters, including detection probability, interference probability, and estimation error. The results show that the proposed sensing techniques can effectively sense the primary receiver and improve the cognitive transmission throughput.

Chengdu, China
Sydney, NSW, Australia
Chengdu, China

Guodong Zhao
Wei Zhang
Shaoqian Li

Contents

Acronyms

AC	Alternating current
AF	Amplify-and-forward
AMC	Adaptive modulation and coding
ATR	Average transmission range
AWGN	Additive white Gaussian noise
CDF	Cumulative distribution function
CLPC	Closed loop power control
CLT	Central limit theorem
CR	Cognitive radio
CT	Cognitive transmitter
DC	Direct current
DTD	Double-threshold detector
ED	Energy detector
EDC	Energy detection comparator
EEECG	End-to-end equivalent channel gain
FDD	Frequency division duplex
FTSEC	Fourier transform of squared envelop of channel
MCS	Modulation and coding scheme
MF	Matched filter
M-BS	Macrocell base station
M-UE	Macrocell user equipment
OTD	One-threshold detector
PDF	Probability density function
PPP	Poisson point process
PR	Primary receiver
PSD	Power spectral density
PT	Primary transmitter
RF	Radio frequency
SH	Spectrum hole
SINR	Signal-to-interference-plus-noise ratio
SNR	Signal-to-noise ratio

S-BS   Small cell base station
TDD   Time division duplex
TDOA   Time difference of arrival
UWB   Ultra-wideband

Chapter 1
Introduction

1.1 Cognitive Radio

In the past decades, the wireless communications experience a fast growth, where the demand of the wireless data grows dramatically. On the other hand, the spectrum resource is limited, which results in spectrum shortage for future wireless devices and services. This is also one of the main bottlenecks in wireless communications.

Cognitive radio (CR) is the solution to deal with the spectrum shortage issue [1, 2]. The basic idea is to let the new wireless devices and services reuse the spectrum bands that have already been allocated to the existing devices and services if the interference to the existing devices can be controlled to an accepted level. In cognitive radio systems, the existing devices are called *primary users* since they have high priority to access the spectrum. In contrast, the new devices are called *cognitive users* or *secondary users* since they have low priority to access the spectrum. Figure 1.1 provides three examples to illustrate how cognitive radio works. In Fig. 1.1a, the cognitive users access the idle time-frequency slots for secondary communication. In Fig. 1.1b, the cognitive users share the same frequency band with primary users, but in different geographic regions. In Fig. 1.1c, the cognitive users coexist with the primary users, but using different spatial directions [3].

In general, the cognitive users can share the same frequency bands with the primary users in two ways [4]. The first one is called *overlay spectrum sharing*, which requires the cognitive users to find the idle frequency bands. Then, cognitive users may access the frequency bands if the primary users are not using it in a certain location and time, e.g., Fig. 1.1a, b. The second one is called *underlay spectrum sharing*, which requires the cognitive users to obtain the cross-channel information. Based on this, the cognitive users may control the interference to the primary users through interference management, e.g., Fig. 1.1c. Therefore, to control the interference to the primary users and establish the cognitive communications, the cognitive users need to obtain the required information of the primary users.

© The Author(s) 2017
G. Zhao et al., *Advanced Sensing Techniques for Cognitive Radio*,
SpringerBriefs in Electrical and Computer Engineering,
DOI 10.1007/978-3-319-42784-3_1

Fig. 1.1 Three examples of
cognitive communication

(a)

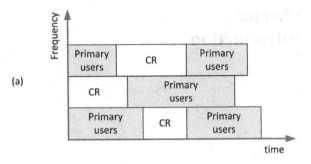

(b)

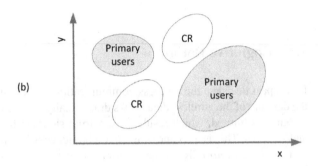

(c)

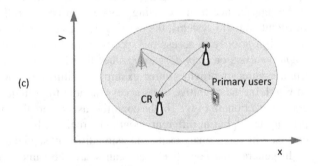

1.2 Spectrum Sensing Techniques

Spectrum sensing is a task that obtains the availability of certain frequency bands,
which is one of the most important components in cognitive radio systems [5].
Figure 1.2a shows the principle of spectrum sensing. In the figure, a *primary
transmitter* (PT) is serving its *receiver* (PR) in a certain coverage. If a cognitive
user intends to concurrently share the same frequency band, it needs to observe
the primary signal coming from the primary transmitter. Once the power of the
observed primary signal is low enough, it indicates that the cognitive user is outside

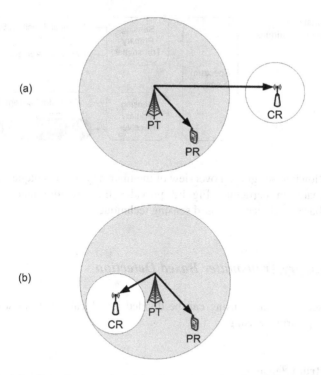

Fig. 1.2 Principle of spectrum sensing: (**a**) primary transmitter based detection; (**b**) primary receiver based detection

the coverage of the primary transmitter. In other words, the primary signal is too weak to establish a communication link and no primary receiver is located in the surrounding region of the cognitive user. As a result, the cognitive user is able to access the same frequency band and does not cause interference to the primary receiver.

The most challenging issue in spectrum sensing is to detect the primary signal in low *signal-to-noise ratio* (SNR) region [6]. This is because the cognitive users are usually located outside the coverage of the primary user. Most of existing works on spectrum sensing aim to detect weak primary signals, i.e., primary transmitter detection.

One the other hand, if we further consider the coverage region of the primary transmitter as shown in Fig. 1.2b, there are still many regions that can be used for cognitive transmission. For example, if the cognitive user is on the left-hand side of the coverage, it can still transmit it own cognitive signal without interfering with the primary receiver. This is because the primary receiver is far away from the cognitive user. Thus, to establish the coexistence of primary and cognitive users, the spectrum sensing needs to detect the nearby primary receiver [7].

Fig. 1.3 Summary of spectrum sensing techniques

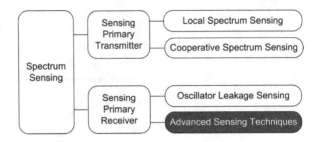

In the following, we give an overview of the main sensing techniques developed in recent years. In particular, Fig. 1.3 provides the structure that consists of transmitter based and receiver based sensing techniques.

1.2.1 Primary Transmitter Based Detection

The PT based spectrum sensing can be divided into local spectrum sensing and cooperative spectrum sensing.

Local Spectrum Sensing

In local spectrum sensing, each cognitive user measures the primary signal and makes a decision by itself. In the following, we introduce some of the most common spectrum sensing techniques.

Energy detection is one of the most widely used spectrum sensing method since it has low complexity and does not require any prior information of the primary signal [8]. In energy detection, the cognitive user measures the energy of the received signal and compares it with a threshold. Once the energy of the received signal is below the threshold, it indicates that the primary transmitter is off or far away. Then, the cognitive user is able to share the frequency band with the primary user.

Matched filter detection is an optimal signal detection method if the cognitive user has a priori information of the primary signal [9]. In matched filter detection, the cognitive user first constructs the primary signal and then conducts the correlation operation. Once the correlation result is above a threshold, the primary transmitter is detected. Since this method uses the information of the primary signal, it has good detection performance. However, it cannot distinguish the signal from the primary device and the interference from other devices, which leads to poor detection performance in the interference scenario.

Cyclostationary-based detection exploits the periodicity of the primary signal [10]. Since the noise is wide-sense stationary with no correlation while the primary signal with a modulation format has cyclic frequency, cyclostationary-based detection is able to differentiate them over some cyclic frequency. Thus, this detection method is suitable for detecting weak signals. In this method, the cognitive user needs to know the modulation information of the primary user, which in practice is not easily accessible.

There are also many other types of detections for spectrum sensing, e.g., waveform based detection, radio identification based detection, wavelet transformed based detection, time-frequency analysis, etc.

Cooperative Spectrum Sensing

In practice, the performance of local spectrum sensing may be degraded due to the uncertainty of the wireless channel, e.g., shadowing and fading. To solve the problem, the cooperative spectrum sensing has been proposed, which is able to enhance the sensing performance through spatial diversity. In cooperative spectrum sensing, the multiple cognitive users are used to obtain the sensing result, which can be divided into two categories: centralized and decentralized cooperative sensing.

In centralized sensing, each cognitive user needs to feedback its measurement to a central unit. Then, the global decision is made based on all measurements. Most of existing works discuss the fusion algorithm [11]. If each sensor makes a local decision and sends only one bit information to the central unit, the global decision is obtained according to some fusion rules, which is referred to as hard decision. On the other hand, if each sensor sends its original measurement to the central unit, the global decision is made by processing of the measurements, which is referred to as soft decision.

In decentralized sensing, each cognitive user shares the local sensing information with each other, but makes its own decision. Thus, it does not require the backbone link between each cognitive user and the central unit [12]. Similarly, in decentralized sensing, different cognitive users may exchange the local measurement or the local decision, which yield different performance. In addition, since the communication link between different cognitive users is not perfect, the feedback error in cooperative sensing was also studied [13]. In summary, compared with local spectrum sensing, the cooperative sensing can achieve better performance. But, the tradeoff between performance and complexity needs to be considered in practical system design.

1.2.2 Primary Receiver Based Detection

Detecting primary receiver is very challenging since the receiver does not transmit signals. The conventional signal detection theory and algorithms cannot be easily applied to the detection of primary receiver. There are very few contributions working in this area.

A method in [14] is to detect the receiver directly via oscillator leakage signals, which is shown in Fig. 1.4. This is because each wireless device has the component

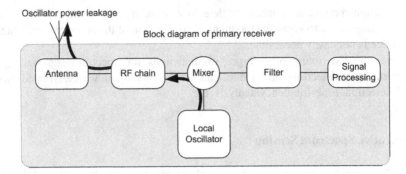

Fig. 1.4 Detecting primary receiver through oscillator power leakage

that converts the received *radio frequency* (RF) signal to the intermediate frequency signal or the base-band signal. During the procedure, the local signal generated by the oscillator inevitably passes through the receive antenna, which is emitted to the air. This is called oscillator power leakage, which occurs in all wireless devices. Therefore, the oscillator power leakage can be used to detect primary receiver. But, the detection algorithm belongs to weak signal detection. Either local or cooperative sensing algorithms can be used.

1.3 Advanced Spectrum Sensing Techniques

In order to obtain high spectrum utilization efficiency, the cognitive and primary users need to coexist in the same frequency band, at the same time, and also in the same geographic region. Hence, the cognitive users need to be capable of detecting its nearby primary receiver or even estimating the cross-channel information between the cognitive transmitter and primary receiver.

This is actually a very challenging task since the primary receiver works passively. Even though the primary receiver may transmit reverse signal back to its transmitter, it is still very hard to conduct the detection and estimation related to primary receiver. This is because in *frequency division duplex* (FDD) systems, the cognitive user may not know which frequency band the primary receiver uses for reverse transmission. In *time division duplex* (TDD) systems, the cognitive user may not know the transmission power of the reverse transmission. As a result, advanced spectrum sensing techniques need to be developed to conduct the detection and estimation that are related to primary receiver.

Fortunately, it is possible for cognitive users to detect the primary receiver and estimate the cross-channel information by exploiting the hidden information of the signal from the primary transmitter. This is because the link adaption has been widely used in existing wireless networks, i.e., the primary networks. This allows the primary transmitter to carry the information related to primary receiver. For example, the transmission power of the primary signal is related to the location of the primary receiver. When the primary receiver is far away from the primary transmitter, the transmission power of the primary signal is high. Otherwise, it is low. Thus, by measuring the power of the primary signal, the cognitive user is able to infer the located region of the primary receiver. Based on the above principle, a region-based sensing technique can be developed to conduct receiver detection.

On the other hand, we may also consider the probing technique to exploit the link adaption between primary transceivers, which can be used to detect the primary receiver and also estimate the cross-channel information between cognitive transmitter and primary receiver. Specifically, if the cognitive transmitter sends some probing signals to the primary receiver, it may affect the primary transmission. Under the link adaption, the primary transmitter automatically adjusts its transmission parameters, e.g., transmission power, *adaptive modulation and coding* (AMC) level, to react the probing. Therefore, by sending probing signal and observing the reaction of the primary signal, the cognitive user is able to detect the primary receiver and estimate the cross-channel information, which is called the jamming-based probing and relay-based probing, respectively.

Table 1.1 compares different advanced sensing techniques. From the table, the region-based sensing and jamming-based probing are both for primary receiver detection. The former is with low implementation complexity and causes no interference to the primary receiver. This is because the region-based method works passively, i.e., the decision is made based on the measured primary signal. In contrast, the jamming-based method needs to conduct the probing, which increases the implementation complexity and causes interference to the primary receiver. But, the detection performance of the jamming-based method is better than the region-based one. If we further consider the relay-based method, it can estimate the cross-channel information and cause low interference, which enable the cognitive user to achieve a satisfactory communication performance. But, this method has high implementation complexity since it requires the cognitive user to be able to conduct the full-duplex relay.

Table 1.1 Comparison of advanced sensing techniques

Technique	Task	Implementation complexity	Interference
Region-based sensing	PR detection	Low	No
Jamming-based probing	PR detection	Medium	High
Relay-based probing	Cross-Ch. estimation	High	Low

1.4   Structure of the Brief

Most of existing sensing techniques detect or estimate the information related to the primary transmitter. However, the information related to primary receiver is more important since the cognitive user needs to manage the interference to primary receiver rather primary transmitter. In this brief, the advanced spectrum sensing techniques are discussed to enable cognitive users to autonomously detect its nearby primary receiver as well as estimate the cross-channel information.

In Chap. 2, we investigate the region-based spectrum sensing technique, where the cognitive femtocell networks are considered. In particular, the receiver detection problem is formulated according to the located geographic regions of primary receiver. Two detectors are designed that use one and two thresholds, respectively. In addition, the performance in terms of the spectrum opportunity for cognitive femtocells is discussed. With the proposed method, the cognitive user is able to detect its nearby active primary receiver.

In Chap. 3, we study the jamming-based probing technique to enable cognitive user to detect its nearby active primary receiver more effectively. The probing principle is discussed and the detection problem is formulated. The probing signal and detection algorithms are designed under both static and dynamic scenarios. The simulation results are provided to show the advantages of the proposed algorithms.

In Chap. 4, we investigate the relay-based probing technique to enable cognitive user to estimate the cross-channel gain between cognitive transmitter and primary receiver. The idea is to use the advanced full-duplex relay technique to reduce the interference to the primary receiver caused by the probing signal. The interference-free probing method is developed. Meanwhile, the cross-channel gain estimator is also obtained. Based on this, the cognitive user is able to conduct the power control to coexist with the primary users.

In Chap. 5, the conclusions are drawn and some future research directions are presented.

References

1. Mitola, J. (2000). Cognitive radio: An integrated agent architecture for software defined radio. Ph.D. dissertation. The Royal Institute of Technology (KTH), Stockholm, Swedem.
2. Haykin, S. (2005). Cognitive radio: Brain-empowered wireless communications. *IEEE Journal on Selected Areas in Communications, 23*(2), 201–220.
3. Zhao, G., Ma, J. Li, Y., Wu, T., Kwon, Y., Soong, A., & Yang, C. (2009). Spatial spectrum holes for cognitive radio with relay-assisted directional transmission. *IEEE Transactions on Wireless Communications, 8*(10), 5270–5279.
4. Zhao, Q. & Sadler, B. M. (2007). A survey of dynamic spectrum access. *IEEE Signal Processing Magazine, 24*, 79–89.
5. Yucek, T. & Arslan, H. (2009). A survey of spectrum sensing algorithms for cognitive radio applications. *IEEE Communications Surveys and Tutorials, 11*(1), 116–130.

6. Tandra, R. & Sahai, A. (2008). SNR walls for signal detection. *IEEE Journal on Selected Topics in Singal Proccessing, 2*(1), 4–17.
7. Akyildiz, I. A., Lee, W. Y., Vuran, M. C., & Mohanty, S. (2006). NeXt generation/dynamic spectrum access/cognitive radio wireless networks: A survey. *Computer Networks, 50,* 2127–2159.
8. Digham, F. F., Alouini, M. S., & Simon, M. K. (2007). On the energy detection of unknown signals over fading channels. *IEEE Transactions on Communications, 55*(1), 21–24.
9. Bhargavi, D. & Murthy, C. (2010). Performance comparison of energy, matched-filter and cyclostationarity-based spectrum sensing. In *Proceedings of IEEE Eleventh International Workshop on Signal Processing Advances in Wireless Communications (SPAWC 2010)*, Marrakech, pp. 1–5, 20–23 June 2010.
10. Sutton, P., Nolan, K., & Doyle, L. (2008). Cyclostationary signatures in practical cognitive radio applications. *IEEE Journal on Selected Areas in Communications, 26*(1), 13–24.
11. Ma, J., Zhao, G., & Li, Y. (2008). Soft combination and detection for cooperative spectrum sensing in cognitive radio networks. *IEEE Transactions on Wireless Communications, 7*(11), 4502–4507.
12. Wang, T., Song, L., Han, Z., & Saad, W. (2014). Distributed cooperative sensing in cognitive radio networks: an overlapping coalition formation approach. *IEEE Transactions on Communications, 62*(9), 3144–3160.
13. Zou, Y., Yao, Y., & Zheng, B. (2011). A selective-relay based cooperative spectrum sensing scheme without dedicated reporting channels in cognitive radio networks. *IEEE Transactions on Wireless Communications, 10*(4), 1188–1198.
14. Wild, B. & Ramchandran, K. (2005). Detecting primary receivers for cognitive radio applications. In *Proceedings of IEEE International Symposium on New Frontiers in Dynamic Spectrum Access Networks (DySPAN 2005)*, Baltimore, pp. 124–130, 8–11 November 2005.

Chapter 2
Region-Based Spectrum Sensing

2.1 Introduction

In cognitive radio systems, spectrum sensing is critical since it determines which kind of spectrum sharing schemes can be used. For example, in cognitive small cell networks, a large number of small cells with short transmission range can share the same spectrum band with a macro cell. To avoid co-channel interference, small cell may access the idle bands of the macro cell, called *overlay spectrum sharing*, which works well in low and medium load scenarios [1, 2], i.e., the macrocell has enough idle bands to accommodate the small cells. However, when the macro cell is with high load, it may have few idle bands. This significantly reduces the transmission opportunities of the small cells.

To deal with the issue, another kind of spectrum sharing, called *underlay spectrum sharing*, is proposed, but it requires the cognitive small cells to control the interference to active *macro cell user equipments* (M-UEs). In the literature, most of existing contributions [3–6] determine an access probability of cognitive small cells to guarantee the outage probability of active M-UEs based on *Poisson point process* (PPP) model. However, those methods can only obtain limited transmission opportunities for cognitive small cells. This is because they are based on the statistic information of the networks and the worst case needs to be considered.

An effective approach to improve the transmission opportunities of the small cells is to allow them to access the busy bands of the macro cell, which requires to sense the location of the active M-UE. If the active M-UE is outside a small cell's coverage, the small cell is allowed to access the M-UE's busy band; if the active M-UE is inside the small cell's coverage, the small cell needs to keep silence to avoid the interference. Therefore, the effective spectrum sharing needs to identify the located region of the M-UE, i.e., whether the active M-UE is inside or outside the coverage of the small cell.

© The Author(s) 2017
G. Zhao et al., *Advanced Sensing Techniques for Cognitive Radio*,
SpringerBriefs in Electrical and Computer Engineering,
DOI 10.1007/978-3-319-42784-3_2

In practice, it is very difficult to detect the located region of the M-UE. This is because most of the existing spectrum sensing methods [7–9] belong to transmitter detection. They cannot be used to detect the M-UE. Specifically, in downlink transmission, since the active M-UE acts as the receiver and does not transmit signals, the small cell can not identify the active M-UE. In uplink transmission, since the small cell does not know the transmission power of the active M-UE, it still can not identify the active M-UE.

In this chapter, a region-based spectrum sensing method is proposed to detect the located region of the active M-UE, where the two-tier cognitive small cell network is considered. In our method, the detection is based on the measured energy of the received signal from a *macro cell base station* (M-BS). We design two detectors with one and two thresholds, respectively, which allow the small cell and the macro cell to simultaneously access the same band. With the proposed method, the *small cell base station* (S-BS) is able to identify the location of the active M-UE and further find the transmission opportunity of the busy band, i.e., the band is being occupied by the active M-UE, but the active M-UE is outside the coverage of the small cell. As a result, the cognitive S-BS can obtain significant transmission opportunities in an "opportunistic" way, i.e., identifying the active M-UE that is outside the coverage of the S-BS and accessing the busy band without interfering with the active M-UE. This is different from conventional overlay methods [7–9] that work in an "on and off" way, i.e., turning on and off the S-BS when a band is idle and busy, respectively. This is also different from conventional underlay methods [3–6] that work in a "blind" way, i.e., accessing the busy band with a certain probability to avoid interference to active M-UEs.

2.2 System Model

Figure 2.1 provides the system model of this chapter. Acting as the primary users, an M-BS serves each M-UE in a certain downlink frequency band. In particular, the M-UEs are uniformly distributed in the coverage of the M-BS with the radius R. On the other hand, acting as a cognitive user, an S-BS inside the M-BS's coverage intends to access the same downlink band being used by the M-UE, where the radius of the S-BS's coverage is r (Here, we use the term "S-BS's coverage" to represent the interference region of the S-BS, then the S-BS does not cause interference to the active M-UE that is outside the S-BS's coverage). Since both the M-UE and S-BS can receive the signal from the M-BS, we will introduce the signal model between the M-BS and M-UE and that between the M-BS and S-BS, respectively. Here, we only consider the three nodes, which do not include the small cell UE.

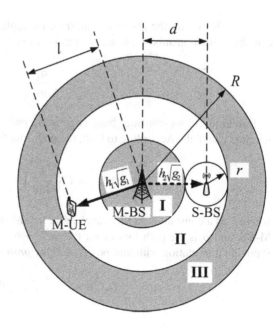

Fig. 2.1 System model (the S-BS that acts as a cognitive user accesses the same frequency band as the M-BS and M-UE that act as the primary users)

2.2.1 Primary Transmission

Denote g_1 as the large-scale path-loss from the M-BS to the M-UE and h_1 as the small-scale fading from the M-BS to the M-UE, then the received signal at the M-UE can be expressed as

$$y_1(k) = h_1 \sqrt{g_1 p} x(k) + n_1(k), \quad 1 \leq k \leq K, \tag{2.1}$$

where $x(k)$ is the transmit signal of the M-BS with the power p, k is the index of the K samples, and $n_1(k)$ is the *additive white Gaussian noise* (AWGN) at the M-UE with zero mean and variance σ_1^2. Then the SNR of the received signal at the M-UE can be expressed as

$$\gamma_1 = \frac{h_1^2 g_1 p}{\sigma_1^2}. \tag{2.2}$$

We assume that the M-BS communicates with the M-UE for guaranteed wireless services. Then, for a certain data rate, the M-BS can automatically adjust the transmission power to meet a target SNR for the specific M-UE, defined by γ_T. This assumption is reasonable since maintaining SNR has been widely applied in practical systems, e.g., the M-BS usually adjusts the transmission power through *close loop power control* (CLPC) or power allocation to provide the required data rate. Based on the above assumption, we replace γ_1 by γ_T in (2.2). Then we can obtain the transmission power of the M-BS by

$$p = \frac{\gamma_T \sigma_1^2}{h_1^2 g_1}. \tag{2.3}$$

Furthermore, we consider both large-scale path-loss and small-scale fading in wireless channels. According to [10], the path-loss follows the model

$$g_1 = C \left(\frac{\lambda}{4\pi l} \right)^\beta, \tag{2.4}$$

where C is a constant, λ is the wavelength, l is the distance between the M-UE and M-BS, and β is the path-loss factor ($\beta = 2 \sim 6$). The small-scale fading follows Rayleigh distribution with unit power, and the *probability density function* (PDF) is

$$f_{h_1}(z) = 2ze^{-z^2}, \quad z \geq 0. \tag{2.5}$$

2.2.2 Cognitive User Measurement

Similarly, denote g_2 as the large-scale path-loss from the M-BS to the S-BS, and h_2 as the small-scale fading from the M-BS to the S-BS, then the received signal at the S-BS can be expressed as

$$y_2(k) = h_2 \sqrt{g_2 p} x(k) + n_2(k), \quad 1 \leq k \leq K, \tag{2.6}$$

where $n_2(k)$ is the AWGN at the S-BS with zero mean and variance σ_2^2. We adopt the same path-loss and small-scale fading models as in (2.4) and (2.5), i.e.,

$$g_2 = C \left(\frac{\lambda}{4\pi d} \right)^\beta \tag{2.7}$$

and

$$f_{h_2}(z) = 2ze^{-z^2}, \quad z \geq 0, \tag{2.8}$$

where d is the distance between the S-BS and M-BS.

Substituting (2.3) into (2.6), we have

$$y_2(k) = \frac{h_2}{h_1} \sqrt{\frac{\gamma_T \sigma_1^2 g_2}{g_1}} x(k) + n_2(k), \quad 1 \leq k \leq K. \tag{2.9}$$

Let

$$\Phi = \frac{g_2}{g_1} \tag{2.10}$$

and

$$\Omega = \frac{h_2{}^2}{h_1{}^2}, \tag{2.11}$$

and assume $\sigma_1^2 = 1$, the received signal at the S-BS in (2.9) can be simplified as

$$y_2(k) = \sqrt{\Phi \Omega \gamma_T} x(k) + n_2(k). \tag{2.12}$$

2.3 Detection Principle

In this section, we introduce the basic principle of our region-based spectrum sensing method that detects the located region of the active M-UE and finds the transmission opportunities for the S-BS. Generally, our region-based method uses the energy of the received signal from the M-BS since the energy carries the location information of the M-UEs. Specifically, as shown in Fig. 2.1, the whole coverage of the M-BS can be divided into three regions according to the location of the S-BS. The S-BS is always in Region II while the M-UEs are randomly located in one of the three regions. If an M-UE is located in Region I or III, the S-BS does not interfere with the M-UE. Otherwise, if an M-UE is in Region II, the S-BS may cause interference to the M-UE. Then we define the two cases as hypotheses $\mathcal{H}_0$ and $\mathcal{H}_1$, respectively, i.e.,

$$\begin{cases} \mathcal{H}_0 : \text{M-UE} \in \text{Regions I or III}, \\ \mathcal{H}_1 : \text{M-UE} \in \text{Region II}. \end{cases} \tag{2.13}$$

For a guaranteed wireless service to the M-UE with a target SNR requirement, the transmission power of the M-BS is mainly determined by the location of the M-UE. If the M-UE is in Region I, which is the center of the M-BS's coverage, the M-BS can meet the M-UE's target SNR with low power; if the M-UE is in Region II, the M-BS needs medium power to satisfy the target SNR; if the M-UE is in Region III, which is the edge of the M-BS's coverage, the M-BS has to use high power for the target SNR. In other words, it establishes the above corresponding relationship between the M-UE's located region and the M-BS's power. Based on the corresponding relationship, the S-BS can distinguish the two hypotheses by measuring the energy of the received signal from the M-BS: if the measured energy is very small or very large, the M-UE is probably in Region I or III, respectively; if the measured energy is medium, the M-UE is probably in Region II. Therefore, the S-BS can use the energy of the received signal from the M-BS as a test statistic to identify the located region of the M-UE.

In the next section, we obtain the distribution of the test statistic and then design the detectors. In particular, we ignore the shadowing in algorithm development because it is very difficult to obtain the closed-from distribution of the test statistic with shadowing. Instead, we will consider it in the simulation section.

2.4   Detector Design

In this section, we first use three steps to obtain the distribution of the test statistic. Based on that, we design a one-threshold detector in Sect. 2.4.5 to maximize the small cell transmission opportunity subject to an interference constraint. Furthermore, we design a double-threshold detector in Sect. 2.4.6, which can achieve better performance.

2.4.1   The Test Statistic

When the S-BS uses K samples to calculate the energy of the received signal from the M-BS, we obtain the following test statistic,

$$E = \sum_{k=1}^{K} y_2^2(k). \tag{2.14}$$

Substituting (2.12) into (2.14), and simplifying the expression, we obtain

$$E = \sum_{k=1}^{K} \Omega \Phi \gamma_T x^2(k) + \sum_{k=1}^{K} 2\sqrt{\Omega \Phi \gamma_T} x(k) n_2(k) + \sum_{k=1}^{K} n_2^2(k). \tag{2.15}$$

In the above expression, Ω, $x(k)$, Φ, and $n_2(k)$ represent four random variables: Ω is determined by the ratio of the power of the two small-scale fadings h_1 and h_2; Φ is determined by the ratio of two path-loss values g_1 and g_2; $x(k)$ is the signal from the M-BS with normalized power; $n_2(k)$ is the AWGN.

It is very difficult to obtain the distribution of the test statistic in (2.15), since it is a combination of the four random variables. To obtain a closed-form expression, we ignore the noise and make the following approximation,

$$E \approx \sum_{k=1}^{K} \Omega \Phi \gamma_T x^2(k) = K \Omega \Phi \gamma_T. \tag{2.16}$$

Now, the energy of the received signal in (2.16) is determined by the two random variables Ω and Φ.

In the following, we first derive the *probability density functions* (PDFs) of Ω and Φ, respectively, and then obtain the *cumulative distribution function* (CDF) of the test statistic E.

2.4.2 Step 1: Obtain the PDF of Ω

Since the small-scale fadings follow independent Rayleigh distribution, the power of them follows exponential distribution, i.e., $h_1^2 \sim e^{-u}$ and $h_2^2 \sim e^{-u}$. According to [11], the CDF of $\Omega = h_2^2/h_1^2$ can be calculated by

$$F_\Omega(\omega) = \int_0^\infty y f_\Omega(y\omega, y) dy = \frac{\omega}{1+\omega}, \tag{2.17}$$

and the PDF of Ω can be obtained by

$$f_\Omega(\omega) = \frac{1}{(1+\omega)^2}, \tag{2.18}$$

which follows F-distribution.

2.4.3 Step 2: Obtain the PDF of Φ

Since the M-UE is uniformly distributed in the coverage of the M-BS, the PDF of the distance between the M-UE and M-BS can be obtained by

$$f_l(l) = \begin{cases} \frac{2l}{R^2-4dr-\varepsilon^2}, & \varepsilon \le l \le d-r \text{ or } d+r \le l \le R \ (\mathscr{H}_0), \\ \frac{2l}{4dr}, & d-r < l < d+r \quad (\mathscr{H}_1), \end{cases} \tag{2.19}$$

where ε is the minimum distance between the M-UE and M-BS. When the path-loss factor is $\beta = 2$, we substitute (2.4) and (2.7) into (2.10), and then obtain

$$\Phi = \frac{g_2}{g_1} = \frac{l^2}{d^2}. \tag{2.20}$$

Given a distance between the S-BS and M-BS, i.e., d, we can obtain the CDF of Φ, i.e.,

$$\begin{aligned} F_\Phi(\phi) &= P\left(\frac{l^2}{d^2} \le \phi\right) \\ &= \begin{cases} \frac{\phi d^2}{R^2-4dr-\varepsilon^2}, & \varepsilon \le l \le d-r \text{ or } d+r \le l \le R \ (\mathscr{H}_0), \\ \frac{\phi d^2}{4dr}, & d-r < l < d+r \quad (\mathscr{H}_1). \end{cases} \end{aligned} \tag{2.21}$$

Then, the PDF of Φ becomes

$$f_\Phi(\phi) = \begin{cases} \frac{d^2}{R^2 - 4dr - \varepsilon^2}, & \mathcal{H}_0, \\ \frac{d^2}{4dr}, & \mathcal{H}_1. \end{cases} \tag{2.22}$$

2.4.4 Step 3: Obtain the CDF of E

As indicated in (2.16), i.e., $E = K\Omega\Phi\gamma_T$, our test statistic E has the same distribution as Ω for a given value of Φ (K and γ_T are constants). Then, we obtain the conditional PDF of E from (2.18) as follows,

$$f_E(\xi|\Phi) = \frac{K\Phi\gamma_T}{(K\Phi\gamma_T + \xi)^2}. \tag{2.23}$$

As a result, the closed-form CDF of E in both $\mathcal{H}_0$ and $\mathcal{H}_1$ can be obtained by

$$\Pr(\xi \le E^*|\mathcal{H}_0) = \int\limits_{M-UE \in \mathrm{I,III}} \int_0^{E^*} f_E(\xi|\Phi)f_\Phi(\phi) d\xi d\phi$$

$$= \int_\varepsilon^{d-r} \int_0^{E^*} \frac{Kl^2 d^2 \gamma_T}{(Kl^2 \gamma_T + d^2\xi)^2} \cdot \frac{2l}{R^2 - 4dr - \varepsilon^2} d\xi dl + \int_{d+r}^R \int_0^{E^*} \frac{Kl^2 d^2 \gamma_T}{(Kl^2 \gamma_T + d^2\xi)^2} \cdot \frac{2l}{R^2 - 4dr - \varepsilon^2} d\xi dl$$

$$= \frac{d^2 E^*}{(R^2 - 4dr - \varepsilon^2) K\gamma_T} \left[\ln\left(\frac{(d-r)^2 K\gamma_T + d^2 E^*}{\varepsilon^2 K\gamma_T + d^2 E^*} \right) + \ln\left(\frac{R^2 K\gamma_T + d^2 E^*}{(d+r)^2 K\gamma_T + d^2 E^*} \right) \right]. \tag{2.24}$$

and

$$\Pr(\xi \le E^*|\mathcal{H}_1) = \int\limits_{M-UE \in \mathrm{II}} \int_0^{E^*} f_E(\xi|\Phi)f_\Phi(\phi) d\xi d\phi$$

$$= \int_{d-r}^{d+r} \int_0^{E^*} \frac{Kl^2 d^2 \gamma_T}{(Kl^2 \gamma_T + d^2\xi)^2} \cdot \frac{2l}{4dr} d\xi dl$$

$$= \frac{d^2 E^*}{4drK\gamma_T} \ln\left(\frac{(d+r)^2 K\gamma_T + d^2 E^*}{(d-r)^2 K\gamma_T + d^2 E^*} \right). \tag{2.25}$$

Figure 2.2 plots the theoretical CDF curves and the corresponding scenarios of the test statistic E based on (2.24) and (2.25), where different distances between the S-BS and M-BS are considered. We also provide simulation curves for comparison, where the system parameters are the same as that in Sect. 2.5. From Fig. 2.2a, when the S-BS is close to the M-BS, i.e., $d = 100$ m, the CDF curves of $\mathcal{H}_0$ are on the right side of the CDF curves of $\mathcal{H}_1$. However, when the S-BS is far away from

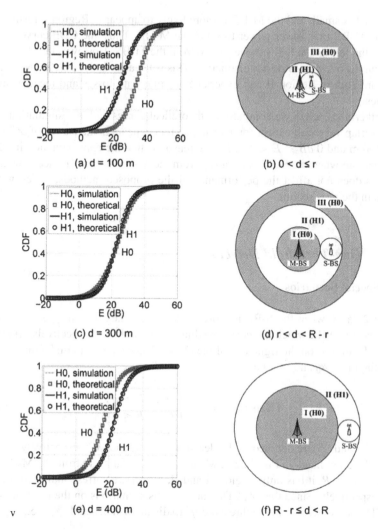

Fig. 2.2 CDF curves of the test statistic E and the corresponding scenarios (d is the distance between S-BS and M-BS, $r = 100$ m, and $R = 500$ m)

the M-BS, i.e., $d = 300$ m and $d = 400$ m in Fig. 2.2c, e, the CDF curves of $\mathcal{H}_0$ dramatically shift to the left side of the curves under $\mathcal{H}_1$. On the other hand, the CDF curves of $\mathcal{H}_1$ slightly shift to the left as the S-BS moves away from the M-BS and the shift is too small to be observed.

This shift is reasonable since the whole coverage of the M-BS is divided into three regions according to the location of the S-BS. When the S-BS is close to the M-BS, Region I is small and Region III is large, and the M-UE is more likely to appear in Region III, resulting in that the M-BS uses higher power to serve the M-UE. Otherwise, when the S-BS is far away from the M-BS, Region I is large and

Region III is small, and the M-UE is more likely to appear in Region I, resulting in that the M-BS uses lower power to serve the M-UE. Thus, the CDF curves of $\mathcal{H}_0$ shift to the left as the S-BS moves away from the M-BS. However, the shift of the CDF curves of $\mathcal{H}_1$ is almost indistinctive. This is because under $\mathcal{H}_1$, the M-UE and S-BS are both in Region II and experience similar path-losses and receive similar energies.

Furthermore, when we compare the theoretical curves with the simulation ones, they overlap very well except for a small gap that appears in the curves of $\mathcal{H}_0$ when $d = 400$ m and $0\ dB < E < 13\ dB$. The reason is that the approximation in (2.16) is inaccurate when the S-BS is far away from the M-BS and experiences low SNR. But this does not affect the performance of the proposed methods, which will be shown in the next section.

2.4.5 One-Threshold Detector

Two Special Scenarios

In Fig. 2.2a, b, when the S-BS is close to the M-BS, i.e., $0 < d \le r$, it has only Regions II and III, which are corresponding to $\mathcal{H}_1$ and $\mathcal{H}_0$, respectively. Since the $\mathcal{H}_0$ CDF curve is on the right side of the $\mathcal{H}_1$ CDF curve, we can find a threshold η' to distinguish $\mathcal{H}_0$ and $\mathcal{H}_1$, i.e.,

$$\text{Decision result} = \begin{cases} \mathcal{D}_0, & E > \eta', \\ \mathcal{D}_1, & E \le \eta', \end{cases} \tag{2.26}$$

where $\mathcal{D}_0$ and $\mathcal{D}_1$ are denoted as the decisions on $\mathcal{H}_0$ and $\mathcal{H}_1$, respectively.

On the other hand, in Fig. 2.2e, f, when the S-BS is far away from the M-BS, i.e., $R - r \le d < R$, it has only Regions I and II, which are corresponding to $\mathcal{H}_0$ and $\mathcal{H}_1$, respectively. Since the $\mathcal{H}_0$ CDF curve in this scenario is on the left side of the $\mathcal{H}_1$ CDF curve, we can find a threshold η'' to distinguish $\mathcal{H}_0$ and $\mathcal{H}_1$, i.e.,

$$\text{Decision result} = \begin{cases} \mathcal{D}_0, & E \le \eta'', \\ \mathcal{D}_1, & E > \eta''. \end{cases} \tag{2.27}$$

The General Scenario

In Fig. 2.2c and d, when the S-BS is in the medium range of the M-BS's coverage, i.e., $r < d < R - r$, it has three regions. Then, the detection becomes complicated because we have only one threshold η. On one hand, if the threshold η is relatively high, the case $E > \eta$ indicates that the M-UE is inside Region III ($\mathcal{H}_0$) and the S-BS is able to access the busy band. In contrast, the case $E \le \eta$ indicates that the M-UE may appear in either Region I ($\mathcal{H}_0$) or II ($\mathcal{H}_1$). Then, the S-BS still has the opportunity (i.e., the probability that the M-UE is in Region I) to assess the busy band. On the other hand, if the threshold η is relatively low, the case $E \leq \eta$ indicates that the M-UE is inside Region I ($\mathcal{H}_0$) and the S-BS is able to access the busy band. In contrast, the case $E > \eta$ indicates that the M-UE may appear in either Region II ($\mathcal{H}_1$) or III ($\mathcal{H}_0$). Then, the S-BS still has the opportunity (i.e., the probability that the M-UE is in Region III) to assess the busy band. As a result, on both sides of the threshold, i.e., $E > \eta$ and $E \leq \eta$, the S-BS has the opportunity to access the busy band.

To maximize the small cell transmission opportunity, it is reasonable to introduce two access probabilities q_1 and q_2 for $E > \eta$ and $E \leq \eta$, respectively, where $0 \leq q_1 \leq 1$ and $0 \leq q_2 \leq 1$. Then, the S-BS can make full use of the transmission opportunities under both cases (i.e., $E > \eta$ and $E \leq \eta$). In particular, with the two access probabilities, the S-BS is able to automatically adjust the decision rule between (2.26) and (2.27), which makes the OTD work in all three scenarios. For example, when $q_1 = 1$ and $q_2 = 0$, the S-BS accesses the busy band if $E > \eta$, which is equivalent to (2.26); when $q_1 = 0$ and $q_2 = 1$, the S-BS accesses the busy band if $E \leq \eta$, which is equivalent to (2.27).

Next, we obtain the optimal access probabilities q_1^* and q_2^* as well as the corresponding optimal threshold η^* so that the small cell transmission opportunity can be maximized.

Calculate the Threshold and Access Probabilities

Based on the above discussion, the probability that the S-BS may access the busy band, i.e., the small cell transmission opportunity, becomes four components, i.e.,

$$P_O = q_1 \left(Pr\{E > \eta | \mathcal{H}_0\} Pr\{\mathcal{H}_0\} + Pr\{E > \eta | \mathcal{H}_1\} Pr\{\mathcal{H}_1\} \right)$$
$$+ q_2 \left(Pr\{E \leq \eta | \mathcal{H}_0\} Pr\{\mathcal{H}_0\} + Pr\{E \leq \eta | \mathcal{H}_1\} Pr\{\mathcal{H}_1\} \right). \quad (2.28)$$

Similarly, the interference probability to the M-UE has two components, i.e.,

$$P_I = \left(q_1 Pr\{E > \eta | \mathcal{H}_1\} Pr\{\mathcal{H}_1\} + q_2 Pr\{E \leq \eta | \mathcal{H}_1\} Pr\{\mathcal{H}_1\} \right) \zeta(d), \quad (2.29)$$

where $\zeta(d) = S_s / S_{\text{II}}$ is the area ratio between the coverage of the S-BS and Region II. The expression of $\zeta(d)$ varies with different ranges of d and we obtain them in [12].

Given an interference constraint I_c, which is the maximum allowable interference probability to the M-UE, the optimal threshold as well as the optimal access probabilities can be obtained by

$$\begin{aligned} & \max_{q_1^*, q_2^*, \eta^*} P_O, \\ & s.t. \ P_I \leq I_c, \ 0 \leq q_1 \leq 1, \ 0 \leq q_2 \leq 1, \ and \ \eta \geq 0. \end{aligned} \quad (2.30)$$

We can substitute (2.28) and (2.29) into (2.30). But, the obtained expression is very complicated since different distance ranges between the S-BS and M-BS need to be considered. Thus, it is very difficult to obtain the closed-form expressions of q_1^*, q_2^*, and η^*. Instead, we will obtain them numerically.

2.4.6 Two-Threshold Detector

In this subsection, we will design another detector using two thresholds. As shown in Fig. 2.2c, when the distance between the S-BS and M-BS is $d = 300$ m, half of the $\mathcal{H}_0$ curve is on the right side of the $\mathcal{H}_1$ curve ($E > 25$ dB), and the other half of the $\mathcal{H}_0$ curve is on the left side of the $\mathcal{H}_1$ curve ($E < 25$ dB). This indicates that the S-BS in such a location should access the busy band if the test statistic of the S-BS is either large enough or small enough. Then, we set two thresholds η_L and η_H to distinguish $\mathcal{H}_0$ and $\mathcal{H}_1$, i.e.,

$$
\begin{cases}
\mathcal{H}_0 : & E \leq \eta_L \ or \ E \geq \eta_H, \\
\mathcal{H}_1 : & \eta_L < E < \eta_H.
\end{cases}
\tag{2.31}
$$

This rule also works for the other two cases when the S-BS is close and far away from the M-BS, respectively, e.g., Fig. 2.2a, e. Then, the small cell transmission opportunity can be expressed as

$$
P_O = Pr\{E \leq \eta_L \ or \ E \geq \eta_H | \mathcal{H}_0\} Pr\{\mathcal{H}_0\}
$$
$$
+ Pr\{E \leq \eta_L \ or \ E \geq \eta_H | \mathcal{H}_1\} Pr\{\mathcal{H}_1\}.
\tag{2.32}
$$

The interference probability to the M-UE can be expressed as

$$
P_I = Pr\{\eta_L < E < \eta_H | \mathcal{H}_1\} Pr\{\mathcal{H}_1\} \zeta(d),
\tag{2.33}
$$

where $\zeta(d)$ is also obtained from [12].

Given an interference constraint I_c, the optimal thresholds that maximize the small cell transmission opportunity can be obtained by

$$
\max_{\eta_L^*, \eta_H^*} P_O,
$$
$$
s.t. \ P_I \leq I_c \ and \ 0 < \eta_L \leq \eta_H.
\tag{2.34}
$$

Similarly, we can also substitute (2.32) and (2.33) into (2.34). However, it is very difficult to obtain the closed-form expression of the optimal thresholds. Therefore, we will calculate them numerically.

2.5 Simulation Results

In this section, we demonstrate the advantages of the proposed detectors, called *one-threshold detector* (OTD) and *double-threshold detector* (DTD). To compare with the conventional transmitter detection, we also provide the performance of the *energy detector* (ED) [13, 14], where the threshold is calculated under Neyman Pearson criteria with 1 % false alarm probability. In the simulation, we assume that the M-UE is uniformly distributed inside the macro cell, where the radius of the macro cell's coverage and that of the small cell's coverage are $R = 500\,\text{m}$ and $r = 100\,\text{m}$, respectively, the target SNR at the M-UE is $\gamma_T = 20\,\text{dB}$, the interference probability constraint is $I_c = 0.01$, the minimum distance between the M-UE and M-BS is $\varepsilon = 36\,\text{m}$ [15], the number of samples is $K = 2$, and $N = 10^4$ Monte Carlo trials are conducted for each simulation curve.

In the following, we first compare the performance of the OTD, the DTD, and the ED. Then, we present their performance as a function of the target SNR and the radius of the S-BS. After that, we demonstrate that our methods work well even if the target SNR of the M-UE is unknown to the S-BS. Finally, we provide the performance of our methods in the case of shadowing.

2.5.1 Comparison of the OTD, the DTD, and the ED

Figure 2.3 compares the transmission opportunities identified by the three detectors, the OTD, the DTD, and the ED, where both simulation and theoretical curves are provided for the OTD and the DTD. From the figure, the simulation and theoretical curves overlap very well, and the proposed OTD and DTD significantly outperform the conventional ED. In particular, the OTD and the DTD have an "U" shape for $d < 500\,\text{m}$ and are with the lowest transmission opportunities when $d \approx 300\,\text{m}$. The reason is that the CDF curves of $\mathcal{H}_0$ and $\mathcal{H}_1$ in Fig. 2.2 overlap when $d \approx 300\,\text{m}$, and then it is hard for the OTD and the DTD to distinguish the two hypotheses. When $d > 500\,\text{m}$ and the S-BS moves out of the coverage of the M-BS, the S-BS may always access the busy band without interfering with the M-BS. Thus, the transmission opportunity approaches 1 for $d > 560\,\text{m}$. Furthermore, when comparing the OTD and the DTD, they have the same performance for most of the S-BS locations except for a small gap when $250\,m < d < 350\,\text{m}$. In this range, the DTD slightly outperforms the OTD.

Figure 2.4 compares the interference probabilities of the OTD, the DTD, and the ED. From the figure, the OTD and the DTD can control their interference under the preset interference probability constraint $I_c = 0.01$ for $d < 500\,\text{m}$. This is because the optimal thresholds of the OTD and the DTD are designed under the interference probability constraint. In addition, the interference probability of the ED is around 0.002 for $d < 500\,\text{m}$, since the ED is designed under the false alarm probability constraint rather than the interference probability constraint. When $d > 500\,\text{m}$

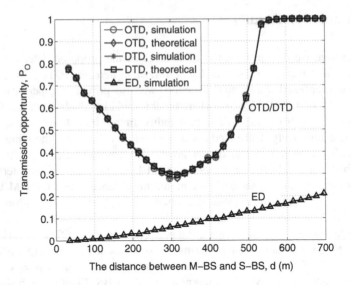

Fig. 2.3 Transmission opportunities of the OTD, the DTD, and the ED

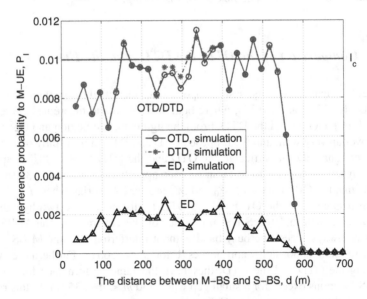

Fig. 2.4 Interference probabilities of the OTD, the DTD, and the ED

and the S-BS moves out of the coverage of the M-BS, there is no M-UE inside the M-BS's coverage and the S-BS causes no interference to the M-UE. Then, the interference probabilities of all three methods approach 0 when $d > 500$ m.

Figures 2.5 and 2.6 provide the optimal threshold and the optimal access probabilities of the OTD, respectively, where the threshold of ED is also provided

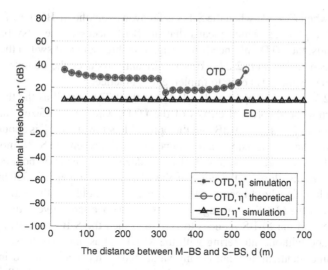

Fig. 2.5 Optimal thresholds of the OTD and the ED

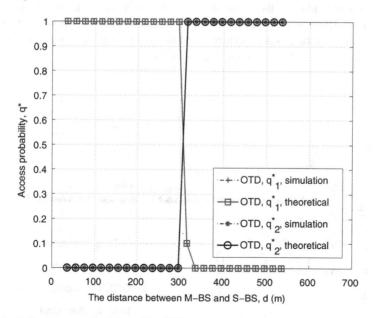

Fig. 2.6 Optimal access probability of the OTD

in Fig. 2.5 for comparison. From Fig. 2.5, the threshold of ED is a constant since it is determined by a fixed false alarm probability and does not change with d. But the optimal threshold of the OTD varies with d since it is calculated based on an interference probability constraint. From Fig. 2.6, when the S-BS is close to the M-BS, i.e., $d < 300$ m, $q_1^* = 1$ and $q_2^* = 0$, which means that the S-BS accesses the busy band when $E > \eta^*$; when the S-BS is far away from the M-BS, i.e., $d > 300$ m, $q_1^* = 0$ and $q_2^* = 1$, which means that the S-BS accesses the busy band when $E < \eta^*$. Thus, the OTD only needs to design one threshold and switch the decision rules between (2.26) and (2.27). This significantly reduces the complexity of the OTD and makes the OTD easy to implement in practice.

Figure 2.7 shows the optimal thresholds of the DTD as well as the threshold of the ED. From the theoretical curves of the DTD, when $d < 250$ m, the optimal high threshold η_H^* is about 30 dB and the optimal low threshold η_L^* is about -80 dB, which means that the decision mainly depends on η_H^*; when the S-BS moves away from the M-BS, i.e., 250 m $< d < 350$ m, the decision depends on both thresholds; when the S-BS is close to the edge of the M-BS's coverage, i.e., 350 m $< d < 500$ m, the low threshold becomes the dominant threshold. As the S-BS keeps on moving out of the M-BS's coverage, $d > 500$ m, the overlap between the coverages of the S-BS and the M-BS reduces to zero, and then the S-BS may always access the busy band without interfering with the M-UE. Thus, the two thresholds of the DTD become identical and can be any value, which has been omitted in Fig. 2.7. Furthermore, the simulation curves match the theoretical ones very well except for the lower threshold η_L^* in the range 36 m $< d < 250$ m and the higher threshold η_H^* in the range 350 m $< d < 500$ m. This is reasonable since the optimal thresholds η_L^* and η_H^* in these ranges approach zero and infinity, respectively. We cannot obtain the theoretical values by conducting the simulation with 10^4 trials.

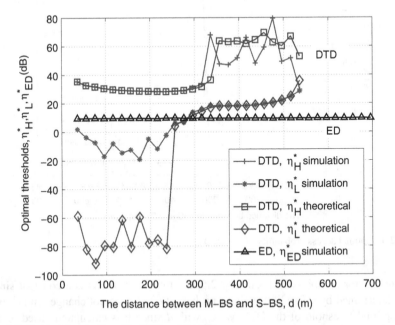

Fig. 2.7 Optimal thresholds of the DTD and the ED

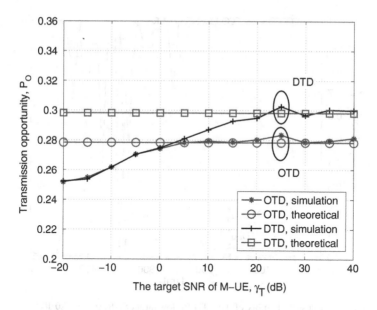

Fig. 2.8 Theoretical and simulation results comparison in the OTD and the DTD

2.5.2   Transmission Opportunities Versus γ_T and r

Figure 2.8 compares the simulation results of the OTD and the DTD with the theoretical ones in terms of the transmission opportunity. Since the largest gap between simulation and theoretical curves in Fig. 2.3 appears at $d \approx 300\,\text{m}$, we evaluate our theoretical results under the worst case of $d = 300\,\text{m}$. From the figure, the simulation and theoretical curves of the OTD do not overlap for $\gamma_T < 5\,\text{dB}$, and these of the DTD do not overlap for $\gamma_T < 15\,\text{dB}$. The reason is that we omit the noise in our derivation, which is not valid in low SNR regions. Fortunately, such an approximation works well in practice, which covers most target SNR values of the M-UE.

Figures 2.9, 2.10 and 2.11 give the two optimal access probabilities of the OTD with low target SNRs. Here, we only provide the simulation results, since the theoretical results do not match the simulation ones for $\gamma_T < 5\,\text{dB}$ as we have discussed. From the three figures, when $d < 300\,\text{m}$, $q_1^* = 1$ and $q_2^* = 0$. But when $d > 300\,\text{m}$, q_1^* and q_2^* become 0 and 1, respectively, as the target SNR γ_T grows from $-20\,\text{dB}$ to $0\,\text{dB}$. This is reasonable and can be explained by our theoretical analysis in [12], i.e., if γ_T is large, the optimal access probabilities are $q_1^* = 1$ and $q_2^* = 0$ for $d < 300\,\text{m}$, and $q_1^* = 0$ and $q_2^* = 1$ for $d > 300\,\text{m}$. In practice, the target SNR of the M-UE is usually larger than $0\,\text{dB}$, i.e., $\gamma_T > 0\,\text{dB}$, then the optimal access probabilities of the OTD can be either 0 or 1, which is also confirmed by Fig. 2.6. Therefore, the proposed OTD is easy to be implemented in practical systems.

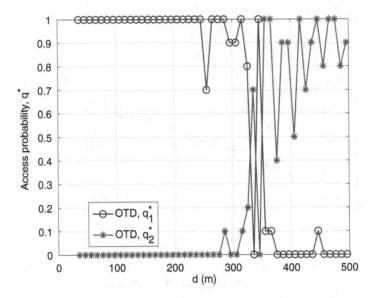

Fig. 2.9 Optimal access probability of the OTD for low target SNR, $\gamma_T = -20\,\text{dB}$

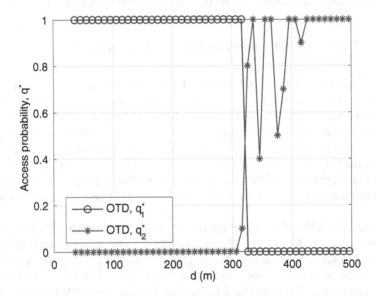

Fig. 2.10 Optimal access probability of the OTD for low target SNR, $\gamma_T = -10\,\text{dB}$

Figure 2.12 shows the transmission opportunities of the OTD and the DTD for different radii of the small cell, where $\gamma_T = 20\,\text{dB}$. Here, we only provide the theoretical results for simplicity since the theoretical and simulation results match very well in high target SNR. From the figure, the transmission opportunities of the OTD and the DTD have the same trend for different distances between the S-BS and

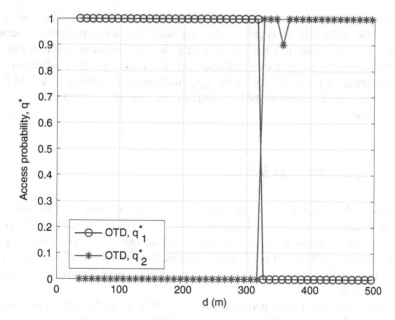

Fig. 2.11 Optimal access probability of the OTD for low target SNRs, $\gamma_T = 0\,\mathrm{dB}$

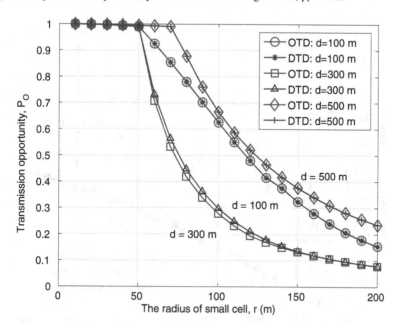

Fig. 2.12 Detection probabilities of the OTD and the DTD versus the radius of small cell

M-BS. Specifically, the transmission opportunity of all curves are equal to 1 for $d <$ 50 m and the S-BS may always access the busy band. The reason is that the coverage of the S-BS is so small that the interference to the M-UE can be ignored. But, as the radius r grows, all the transmission opportunities decrease. This is because the large coverage of the S-BS results in high interference probability to the M-UE. To meet the interference probability constraint, the transmission opportunity of the S-BS reduces.

2.5.3 γ_T Unknown to S-BS

In practice, the target SNR of the M-UE varies with different services and may even be unknown to the S-BS. Thus, we assume that the target SNR is uniformly distributed in a reasonable range, e.g., between 5 dB and 20 dB. Then, the CDF of the test statistic in Sect. 2.4 needs to incorporate the distribution of γ_T. Since it is very difficult to obtain the closed-form CDF in this case, we will provide simulation results instead.

Figure 2.13 compares the performance of the OTD, the DTD, and the ED, where γ_T is uniformly distributed between 5 dB and 20 dB. From the figure, the proposed

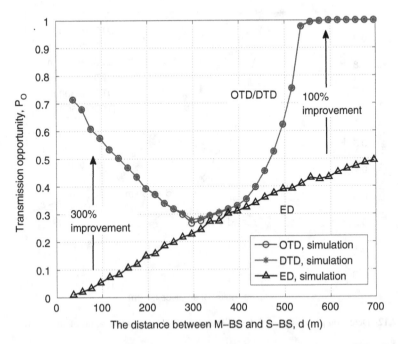

Fig. 2.13 Transmission opportunities of the OTD, the DTD, and the ED (the target SNR of M-UE is unknown to S-BS)

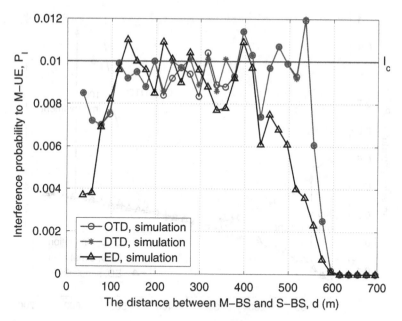

Fig. 2.14 Interference probabilities of the OTD, the DTD, and the ED (the target SNR of M-UE is unknown to S-BS)

the OTD and the DTD obtain almost the same performance. They achieve about 100 % to 300 % more transmission opportunities in average than the conventional ED. Figure 2.14 provides the corresponding interference probabilities of the OTD, the DTD, and the ED. From the figure, the three methods have similar interference to the M-UE, i.e., the interference probabilities are around 0.01 for $d < 500$ m, and approach 0 for $d > 500$ m.

2.5.4   The Case of Shadowing

In this subsection, we provide the performance of the three detectors in the case of shadowing, i.e., path-loss, shadowing, and multi-path fading are all considered. In particular, the shadowing coefficient follows log-normal distribution with the standard variation of 12 dB (We actually provide the worst case performance since the standard variation of shadowing is usually between 4 dB and 12 dB [10]). All simulation parameters except $K = 16$ are the same as those in Figs. 2.13 and 2.14. Here, the reason that the number of samples K is set to 16 is to let the ED has similar interference probability with the OTD and the DTD. This is because the OTD and the DTD are designed under the interference probability constraint. When we consider shadowing, they can automatically adjust the thresholds to reach the preset interference probability $I_c = 0.01$. However, the ED is designed under the

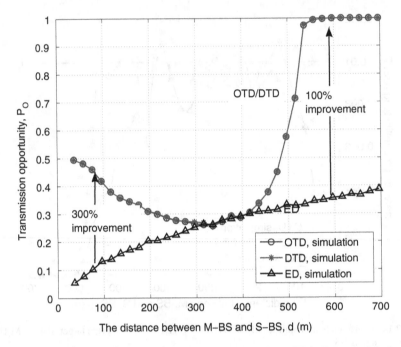

Fig. 2.15 Transmission opportunities of the OTD, the DTD, and the ED in the case of shadowing (the target SNR of M-UE is unknown to S-BS)

false alarm probability. When we consider shadowing, the ED does not change the threshold, which raises the interference probability. Thus, we adjust the number of samples K to raise the threshold and make the ED have similar interference probability with the OTD and the DTD. This allows us to make a comparison of different detectors in terms of the transmission opportunity.

Figure 2.15 provides the transmission opportunities of the three detectors. From the figure, we observe the same trend as in Fig. 2.13, but with only slight performance loss at all detectors. Figure 2.16 shows the corresponding interference probabilities of all three detectors. From the figure, we observe almost the same curves as in Fig. 2.14. Therefore, in the case of shadowing, the proposed OTD and DTD can still achieve about 100 % to 300 % more transmission opportunities in average than the conventional ED.

2.6 Summary

In this chapter, a region-based spectrum sensing method was proposed to enable a cognitive small cell to identify the location of an active user that is being served by a macro cell. Then, the small cell may access the busy band if the active

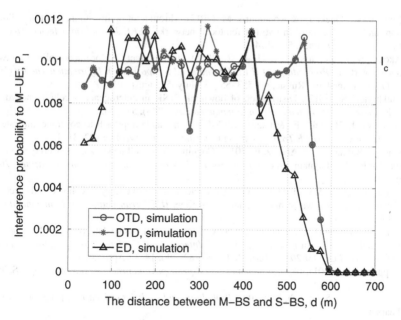

Fig. 2.16 Interference probabilities of the OTD, the DTD, and the ED in the case of shadowing (the target SNR of M-UE is unknown to S-BS)

user is outside the coverage of the small cell, and achieve more transmission opportunities. We suggested that the small cells detect the active user by using the received energy from the macro cell as the test statistic. Then we designed two detectors with one and two thresholds, called OTD and DTD, respectively. Meanwhile, as the conventional energy detector, neither the OTD nor the DTD requires any prior information of the macro cell's signal. Our results indicated that under the same interference probability to the active user, the proposed OTD and DTD achieve about 100 % to 300 % more transmission opportunities in average than the conventional energy detector. This is because our region-based detectors exploit the active user's location information, which is carried by the power of the macro cell's signal for guaranteed services with target SNR requirement. Therefore, the proposed OTD and DTD are suitable for the macro cell with guaranteed services in future wireless communication systems.

References

1. Lopez, D., Valcarce, A., Roche, G., & Zhang, J. (2009). OFDMA femtocells: A road map on interference avoidance. *IEEE Communications Magazine, 47*(9), 41–48.
2. Lopez, D., Guvenc, I., Roche, G., Kountouris, M., Quek, T., & Zhang, J. (2011). Enhanced inter-cell interference coordination challenges in heterogeneous networks. *IEEE Wireless Communications, 18*(3), 22–30.

3. Cheung, W., Quek, T., & Kountouris, M. (2012). Throughput optimization, spectrum allocation, and access control in two-tier femtocell networks. *IEEE Journal on Selected Areas in Communications, 30*(3), 561–574.
4. Akoum, S., Kountouris, M., & Heath, R. (2011). On imperfect CSI for the downlink of a two-tier network. In *Proceedings of IEEE International Symposium on Information Theory (ISIT 2011)*, St. Petersburg, Russia, pp. 553–557, 31 July–5 August 2011.
5. Mukherjee, S. (2012). Distribution of downlink SINR in heterogeneous cellular networks. *IEEE Journal on Selected Areas in Communications, 30*(3), 575–585.
6. Andrews, J., Baccelli, F., & Ganti, R. (2011). A tractable approach to coverage and rate in cellular networks. *IEEE Transactions on Communications, 59*(11), 3122–3134.
7. Gur, G., Bayhan, S., & Alagoz, F. (2010). Cognitive femtocell networks: An overlay architecture for localized dynamic spectrum access. *IEEE Wireless Communications Magazine, 17*(4), 62–70.
8. Wu, J., Wang, C., & Wang, T. (2011). Performance analysis of energy detection based spectrum sensing with unknown primary signal arrival time. *IEEE Transactions on Communications, 59*(7), 1779–1784.
9. Wang, W., Zhang, J. & Zhang, Q. (2013). Cooperative cell outage detection in self-organizing femtocell networks. In *Proceedings of IEEE International Conference on Computer Communications (INFOCOM 2013)*, Turin, Italy, pp. 782–790, 14–17 April 2013.
10. Rappaport, T. (2001). *Wireless communication-principle and practice (2nd ed.)*. Upper Saddle River: Prentice Hall.
11. DeGroot, M. & Schervish, M. (2011). *Probability and statistics (4th ed.)*. Upper Saddle River: Pearson.
12. Li, L., Zhao, G., & Zhou, X. (2015). Enhancing small cell transmission opportunity through passive receiver detection in two-tier heterogeneous networks. *IEEE Transactions on Signal Processing, 63*(13), 3461–3473.
13. Trees, H. (2001). *Detection, estimation, and modulation theory (Part I)*. Hoboken: Wiley-Interscience.
14. Digham, F., Alouini, M., & Simon, M. (2007). On the energy detection of unknown signals over fading channels. *IEEE Transactions on Communications, 55*(1), 21–24.
15. 3GPP TR 25.814. (2006). *Physical layer aspects for evolved universal terrestrial radio access (UTRA)*, 2006.

Chapter 3
Jamming-Based Probing for Spectrum Sensing

3.1 Introduction

In *cognitive radio* (CR), unlicensed users are able to access the frequency band
that belong to primary users as long as they do not interfere with the primary users
[1–5]. Thus spectrum sensing that finds spectrum opportunities in primary networks
is critical to enable CR users to work on a non-interference basis.

Spectrum holes (SHs) [2, 3] describe spectrum opportunities and are defined as
the vacant spectrum bands that can be accessed by CR users in a certain time and
at a certain geographic location. By detecting *primary transmitters* (PTs), CR users
can identify SHs and conduct overlay spectrum sharing [5–17]. On the other hand,
the spectrum opportunity can also be exploited by underlay spectrum sharing as
in [18, 19]. That is, CR users may coexist with primary users at the same time in
the same spectrum band and geographic area if the interference generated by CR
users is below a tolerant threshold. For example, a CR transmitter can spread its
power over a wide band by spread spectrum [20] or ultra-wideband (UWB) [21]
techniques. However, the transmit power of such a scheme has to be conservative
and it is limited to short range applications.

Since the ultimate goal of spectrum sensing is to avoid interfering with primary
receivers (PRs), it is more important to detect PRs directly. Based on the PR
detection results, CR users can budget their transmit power and exploit the spectrum
opportunities more efficiently. In [22], a direct sensing method to detect PRs has
been developed by exploiting local oscillator leakage emitted by *radio-frequency*
(RF) front end of a PR. However, this approach has poor performance because the
leakage signal is too weak to be detected.

© The Author(s) 2017
G. Zhao et al., *Advanced Sensing Techniques for Cognitive Radio*,
SpringerBriefs in Electrical and Computer Engineering,
DOI 10.1007/978-3-319-42784-3_3

In this chapter, a jamming-based probing method is proposed to effectively detect PRs. Different from traditional spectrum sensing methods [6–18] that detect SHs by *listening* to primary signals, the proposed method finds SHs by probing the *close-loop power control* (CLPC) of the primary link, i.e., sending a jamming signal and observing possible corresponding power fluctuation of the primary signal. As a result, the performance of CR system can be significantly enhanced.

3.2   System Model

Figure 3.1 shows an example of a CR application in a cellular primary system. It is assumed that a PT and a PR-A are communicating with each other by *frequency division duplex* (FDD). From the figure, Channels 1 and 2 are used for downlink and uplink transmissions respectively between the PT and PR-A. A CR transmitter is in the coverage of the PT and intends to communicate with a CR receiver by Channel 1. If the coverage of the CR communication is the shaded area in the figure, the CR user may still use Channel 1 even though the channel is already occupied by the primary user because PR-A is beyond the range of the CR communication. In this scenario, PR-A is referred to a *coexistence terminal* since it is out of CR's coverage and may coexist with CR users. In contrast, PR-B in the figure is called an *interference terminal*. Assume that PR-B uses Channels 3 and 4 for downlink and uplink transmissions, respectively. Since PR-B is within the coverage of the CR transmitter, the CR user is not allowed to use Channel 3 that is assigned to PR-B. The purpose of the jamming-based probing method is to distinguish coexistence terminals from the interference ones. Once a coexistence terminal is identified, the CR user is able to communicate over its downlink channel.

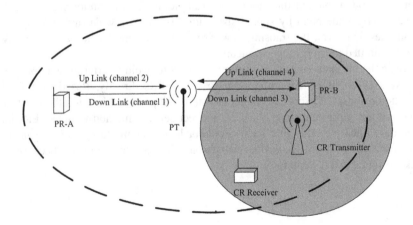

Fig. 3.1 CR system setup in a cellular primary system

Denote P_r, P_i, and P_n as the power of the desired signal, the interference power from CR users, and the noise power at a PR, respectively. Then the *signal-to-interference-plus-noise ratio* (SINR) is given by

$$\gamma = \frac{P_r}{P_i + P_n}. \tag{3.1}$$

In order to maintain an desired signal quality at the PR, CLPC will adjust the PT's transmit power to compensate the power loss caused by characteristics of wireless channels, such as path-loss and shadowing. Denote G as the channel gain between the PT and the PR, and P_t as the transmit power of the PT, then

$$P_r = G \cdot P_t. \tag{3.2}$$

From (3.1) and (3.2), the transmit power of the PT under CLPC can be obtained by

$$P_t = \frac{\gamma_T (P_i + P_n)}{G}, \tag{3.3}$$

where γ_T denotes the target SINR for the desired signal quality. Usually, quantized information is used for CLPC feedbacks in practical wireless systems [23]. To facilitate analysis, an ideal feedback is assumed in CLPC, where there are no quantization and transmission errors.

3.3 Detection Principle

As indicated before, the PT's transmit power with CLPC is adjusted according to the interference power at the PR to maintain the quality of the received signal. When a CR user is close to a PR, the jamming signal changes the interference environment of the PR and therefore changes the PT's transmit power accordingly. If the power fluctuation of the PT signal is detected as a response to the jamming signal, then, with a high probability, the CR transmission is causing interference to the PR. Thus it needs to vacate the spectrum band as soon as possible. This is referred to as Hypothesis $\mathcal{H}_B$, which denotes the channel as busy. On the other hand, when the CR user is far away from the PR, the jamming signal does not trigger the primary's CLPC and the CR transmission may coexist with the primary transmission at the same time and in the same geographic area. This is referred to as Hypothesis $\mathcal{H}_I$, which means that the channel is idle and can be accessed by CR users. In summary, by sending a jamming signal and observing whether the transmit power of the PT changes accordingly, a CR user can determine whether a PR is within its interference region.

In the system, a whole sensing frame is divided into several observation blocks. The channel gains are constant during each block, but may vary from block to block.

In the following, the power of the received signal in a single block is first derived, where the index of the block is omitted for simplicity. Then, the power in the whole sensing frame is considered to formulate the PR detection problem.

As an extension, the *modulation and coding scheme* (MCS) can also be exploited in our jamming-based probing method. For example, when the jamming signal changes the interference environment of a nearby PR, the PT may adjust the MCS to maintain the reliability of the primary communication link. Then CR users can identify the nearby PR by detecting such responses using MCS classification approaches proposed in [24] and [25].

3.3.1 Power of the Received Signal in a Single Block

Denote $s(t)$ as the transmit signal of the PT with the power P_t, then the received signal at the CR user can be expressed as

$$y(t) = h\sqrt{P_t}s(t) + n(t),   \tag{3.4}$$

where h denotes the channel coefficient between the PT and the CR user, and $n(t)$ represents *additive white Gaussian noise* (AWGN) with zero mean and variance σ_n^2. According to [26], the energy of the received signal, denoted as Y', follows a non-central chi-square (χ^2) distribution with M degrees of freedom and a non-central parameter $M\mu$, i.e.,

$$Y' \sim \chi_M^2(M\mu),   \tag{3.5}$$

where

$$\mu = \frac{\sum_{k=0}^{M-1}\left[h\sqrt{P_t}s\left(\frac{kT}{M}\right)\right]^2}{M\sigma_n^2}.   \tag{3.6}$$

It is the instantaneous *signal-to-noise-ratio* (SNR) in the current observation block. According to the *central limit theory* (CLT), Y' is approximately Gaussian when M is large, i.e.,

$$Y' \sim \mathcal{N}[M(1+\mu), 2M(1+2\mu)],   \tag{3.7}$$

where $\mathcal{N}[a, b^2]$ represents Gaussian distribution with mean a and variance b^2. Then the power of the received signal in the current observation block can be obtained by $Y = \frac{Y'}{M}$. Obviously,

$$Y \sim \mathcal{N}[(1+\mu), \frac{2}{M}(1+2\mu)].   \tag{3.8}$$

3.3.2 Power of the Received Signal in Multiple Blocks

In the entire sensing frame, the CR user calculates the power of the received signal within every block, and each of them follows a Gaussian distribution with mean $1 + \mu(l)$ and variance $\frac{2}{M}(1 + 2\mu(l))$, where l is the index of the observation block, the mean $1 + \mu(l)$ represents the expected power related to wireless channels and CLPC, and variance $\frac{2}{M}(1 + 2\mu(l))$ reflects the uncertainty of the received power caused by the noise.

Denote

$$\eta(l) = Y(l) - (1 + \mu(l)). \tag{3.9}$$

Then,

$$\eta(l) \sim \mathcal{N}[0, \frac{2}{M}(1 + 2\mu(l))], \tag{3.10}$$

and

$$Y(l) = 1 + \mu(l) + \eta(l), \tag{3.11}$$

where $1 + \mu(l)$ can be regarded as the signal component, $\eta(l)$ as the noise component, and $\mu(l) = |h(l)|^2 P_t(l)$. Here, $\sigma_n^2 = 1$ in (3.6).

The power of the jamming signal is divided into *direct current* (DC) component and *alternating current* (AC) component. The DC component denotes the average power of the jamming signal and equals to 1 for simplicity while the AC component represents the power fluctuation of the jamming signal and is denoted as $x(l)$. Then the jamming signal can be expressed as $(1 + x(l))$. Since only the AC component is focused for PR detection, the term *jamming signal* will represent $x(l)$ in the rest of this section for simplicity. Then the interference power at the PR can be expressed as

$$P_i(l) = \alpha(1 + x(l)), \tag{3.12}$$

where α reflects the interference strength. The larger α is, the more interference to PRs there is and vice versa. Assume $P_n = 1$, substituting (3.12) into (3.3) leads to the transmit power of a PT using CLPC as follows,

$$P_t(l) = p[1 + \alpha(1 + x(l))], \tag{3.13}$$

where $p = \frac{\gamma_T}{G}$. From (3.11) and (3.13), the power of the received signal at the CR user can be expressed as

$$Y(l) = p(1 + \alpha)|h(l)|^2 + p\alpha|h(l)|^2 \cdot x(l) + 1 + \eta(l), \tag{3.14}$$

which indicates that $Y(l)$ will be affected by the jamming signal, $x(l)$, if the CR user is close to the PR. Thus Hypotheses $\mathcal{H}_B$ and $\mathcal{H}_I$ can be distinguished by detecting whether $x(l)$ is present or absent in $Y(l)$. Perform *Fourier* transformation to (3.14), we have

$$\tilde{Y}(f) = p(1 + \alpha)\tilde{H}(f) + p\alpha[\tilde{H}(f) \otimes \tilde{X}(f)] + \delta(f) + \tilde{P}_n(l), \qquad (3.15)$$

where $\otimes$ denotes the convolution operation, $\tilde{Y}(f)$, $\tilde{H}(f)$, $\tilde{X}(f)$, and $\tilde{P}_n(f)$ represent the *Fourier* transforms of $Y(l)$, $|h(l)|^2$, $x(l)$, and $\eta(l)$, respectively.

In (3.15), the received signal consists of four components in frequency domain if $\mathcal{H}_B$ is true. The first one is the *Fourier transform of squared envelop of channel* (FTSEC) and it depends on wireless channels, the second one is the convolution of FTSEC and the jamming signal, the third one is a DC signal which is a constant and can be subtracted from the received signal, and the last one is the noise component. On the other hand, if $\mathcal{H}_I$ is true, only the first and the last terms exist in (3.15).

To summarize, the power of the received signal from a PT is the signal of interest. By detecting whether it is adjusted according to the jamming signal, the CR user can identify nearby PRs or access the spectrum band if the interference to PRs is below a tolerant threshold.

3.4 Probing and Detection Algorithms

As indicated in [23], the maximum component of *power spectrum density* (PSD) of the channel envelop is $2f_m$, where f_m represents the maximum doppler frequency in wireless channel. Thus the frequency components of $x(l)$ should be larger than $2f_m$. Otherwise, spectrum overlaps between the jamming signal and the variation of the wireless channels will confuse the CR user and degrade the detection performance. On the other hand, the frequency components of $x(l)$ have to be limited to a certain value so that the CLPC is able to respond them. That is because CLPC is to compensate SINR losses caused by shadowing or path-loss in wireless channels [27], which do not change rapidly. Based on the above two constraints, slow fading channels and fast fading channels are considered. In slow fading channels, f_m is much smaller than the frequencies of the jamming signal, while in fast fading channels, f_m is much larger than the frequencies of the jamming signal.

3.4.1 Static Scenario

As discussed before, the PR can be detected by observing the power variation of the received primary signal. Specifically, the CR user needs to identify whether or not such power variations are caused by its jamming signal under fluctuations of wireless channels. To distinguish two signals with different frequency bands,

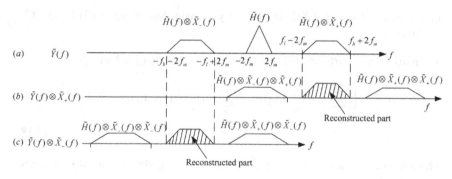

Fig. 3.2 Principle of the MF detector

we design the detector in frequency domain. We will consider a *Fourier* transform approach to design a jamming signal and the corresponding detection algorithm. Assume that the jamming signal, $x(l)$, is band-pass, where f_l and f_h represent its lowest and highest frequency components, respectively. Since $x(f)$ is non-negative, its *Fourier* transform $\tilde{X}(f)$ is symmetric. Define

$$\tilde{X}(f) = \tilde{X}_-(f) + \tilde{X}_+(f), \tag{3.16}$$

where $\tilde{X}_-(f) = \tilde{X}_+(-f)$, and $\tilde{X}_-(f)$ and $\tilde{X}_+(f)$ represent the negative and the positive components of $\tilde{X}(f)$, respectively. Figure 3.2a shows the *Fourier* transform of $Y(l)$ if $\mathcal{H}_B$ is true, where we omit the DC and noise components for simplicity. As shown in the figure, if $f_l - 2f_m > 2f_m$ or $f_l > 4f_m$, $\tilde{Y}(f)$ consists of three isolated components, $\tilde{H}(f) \otimes \tilde{X}_-(f)$, $\tilde{H}(f)$, and $\tilde{H}(f) \otimes \tilde{X}_+(f)$. On the other hand, if $\mathcal{H}_I$ is true and $\tilde{X}(l) = 0$, there is only $\tilde{H}(f)$ component left in $\tilde{Y}(f)$. Therefore, by detecting $\tilde{H}(f) \otimes \tilde{X}_-(f)$ and $\tilde{H}(f) \otimes \tilde{X}_+(f)$ components, the CR user can distinguish $\mathcal{H}_I$ and $\mathcal{H}_B$. Equation (3.10) indicates that the variance of the noise varies with $\mu(l)$, i.e., the noise is non-stationary. To facilitate analysis, we still assume it is stationary and apply *matched filter* (MF) for the detection of $\tilde{X}(f)$. Again as shown in Fig. 3.2a, $\tilde{H}(f)$ is a fraction of $\tilde{Y}(f)$, where $-2f_m < f < 2f_m$. Then the CR user can obtain $\tilde{H}(f)$ through the Fourier transformation of the envelop of the received signal no matter whether or not CLPC is triggered by the jamming signal. Therefore, $\tilde{H}(f) \otimes \tilde{X}_-(f)$ and $\tilde{H}(f) \otimes \tilde{X}_+(f)$ can be reconstructed by $\tilde{Y}(f) \otimes \tilde{X}_-(f)$ and $\tilde{Y}(f) \otimes \tilde{X}_+(f)$, as illustrated in Fig. 3.2b, c, respectively. As in the figure, the shaded components of the spectrum are the reconstructed parts. Then the test statistics of the MF detector can be obtained by performing correlation operation between the received signal, $\tilde{Y}(f)$, and the reconstructed signal, $\tilde{Y}(f) \otimes \tilde{X}(f)$, in frequency domain, i.e.,

$$D = \sum \tilde{Y}(f) \cdot [\tilde{Y}(f) \otimes \tilde{X}(f)]. \tag{3.17}$$

In slow fading channels, we assume $f_l \gg f_m$ and $f_h \gg f_m$ so that there is no interference when performing correlation operations among different isolated components of the signal $Y(l)$ in frequency domain. Substitute (3.15) into (3.17), we obtain

$$D = p^2\alpha(1+\alpha)\sum\{[\tilde{H}(f) \otimes \tilde{X}(f)]^2 + \tilde{H}(f) \cdot [\tilde{H}(f) \otimes \tilde{X}(f) \otimes \tilde{X}(f)]\} + \tilde{N}(f)$$

$$= 2p^2\alpha(1+\alpha)\left\{\sum_{f_l-2f_m}^{f_h+2f_m}[\tilde{H}(f) \otimes \tilde{X}_+(f)]^2 + \sum_{-2f_m}^{2f_m}\tilde{H}(f) \cdot [\tilde{H}(f) \otimes \tilde{X}_+(f) \otimes \tilde{X}_-(-f)]\right\} + \tilde{N}(f),$$

$$(3.18)$$

where $\tilde{N}(f)$ is the noise component. Then the jamming signal of a CR user can be designed as follows,

$$\max_{\tilde{X}(f)}(D) \tag{3.19}$$

$$s.t. \sum_{f=f_l}^{f_h} \tilde{X}(f) = \text{constant},$$

$$f_h \leq f_{CLPC},$$

where f_{CLPC} is the highest frequency that CLPC is able to react.

In (3.19), the jamming signal can be obtained in frequency domain by maximizing the output of the MF detector and it also needs to satisfy the two constraints, i.e., the summation of frequency components of the jamming signal is constrained so that it will not cause the harmful interference to PRs; the highest frequency component of $\tilde{X}(f)$ is smaller or equals to f_{CLPC} so that it can be responded by CLPC. As shown in Fig. 3.2, with a proper jamming signal, the original signal is able to be matched perfectly. It is very hard to obtain a closed-form expression of the jamming signal from (3.19). Therefore, from practical considerations, we apply a sinusoid signal with frequency f_x as our jamming signal, which facilitates the detection as well, and thus $f_l = f_h = f_x$. Substitute $\tilde{X}_+(f) = \delta(f - f_x)$ and $\tilde{X}_-(f) = \delta(f + f_x)$ into (3.18), then we obtain the test statistics as

$$D = 4p^2\alpha(1+\alpha)\sum_{-2f_m}^{2f_m}\tilde{H}^2(f) + \tilde{N}(f) \quad \text{if } f_x > 4f_m. \tag{3.20}$$

It can be found that the effect of the sinusoid jamming signal is equivalent to shifting a FTSEC, $\tilde{H}(f)$, to a high frequency region. When $f_x > 4f_m$, there is no overlap between $\tilde{H}(f)$ and its shifted version $\tilde{H}(f \pm f_x)$. Then the jamming signal can be easily identified by the MF detector.

3.4.2 Dynamic Scenario

In this case, $f_{CLPC} \geq f_x > 4f_m$ does not always hold. As will be shown in the next subsection, the performance of the MF detector decreases dramatically when $f_x \leq 4f_m$. To enable our proposed scheme in this scenario, we develop a jamming signal and the corresponding detector that are suitable for fast fading channels.

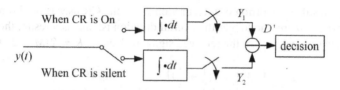

Fig. 3.3 Diagram of the EDC detector

Rectangle pulses can be applied as a jamming signal and can be implemented by turning a CR signal on and off alternately. In particular at the CR user, the power fluctuation caused by the variation of channel gains can be averaged in a fast fading channel and it leads to good detection performance in this scenario. Figure 3.3 shows the diagram of the detector. It calculates the power of the received signal during CR's on and off periods separately and then compares the difference between them to get the test statistics for decision. Thus nearby PRs can be detected by the CR user and such a detector is named as an *energy detection comparator* (EDC).

Assume that the overall durations of CR's on and off periods are T_1 and T_2, respectively, and T is the duration of one observation block. Let $a = \frac{T_1}{T}$ and $b = \frac{T_2}{T}$. In duration T, we denote Y_1 and Y_2 as the equivalent energies when CR is on and off, respectively. Then, it can be shown that

$$Y_1 \sim \mathcal{N}\left[M + \gamma_{CR}, \frac{2(M + 2\gamma_{CR})}{a}\right], \tag{3.21}$$

and

$$Y_2 \sim \mathcal{N}\left[M + \gamma_{CR} - \theta, \frac{2(M + 2(\gamma_{CR} - \theta))}{b}\right], \tag{3.22}$$

where M is the number of samples in a unit duration, γ_{CR} denotes the SNR at the CR user, and θ represents the transmit SNR adjustment when the jamming signal turns from on to off. In particular, θ represents the power variation of the PT's signal due to the CLPC triggered by the jamming signal. Thus if there is a PR nearby, the expected values of Y_1 and Y_2 may be different, otherwise, they should be the same. Therefore, the PR can be detected based on the test statistics, $D' = Y_2 - Y_1$, which follows a Gaussian distribution, that is,

$$D' \sim \mathcal{N}\left[\theta, \frac{2(M + 2\gamma_{CR})}{a} + \frac{2(M + 2\gamma_{CR} - 2\theta)}{b}\right]. \tag{3.23}$$

Define the duty cycle of the jamming signal as

$$\nu = \frac{a}{a + b}, \tag{3.24}$$

which denotes the proportion of the duration when the CR user is on. From (3.23), the larger a and b are, the longer the overall duration is, and as a result, the smaller the variance of the power of the received signal is. Denote $K = 2(M + 2\gamma)$, then

$$D' \sim \begin{cases} \mathcal{N}\left[0, K(\frac{1}{a} + \frac{1}{b})\right], & \mathcal{H}_I \ (\theta = 0), \\ \mathcal{N}\left[\theta, K(\frac{1}{a} + \frac{1}{b}) - \frac{4\theta}{b}\right], & \mathcal{H}_B \ (\theta > 0). \end{cases} \tag{3.25}$$

Given the threshold of the detector, λ, the probabilities of false alarm and miss detection can be expressed as

$$P_f = Q\left(\frac{\lambda}{\sqrt{K(\frac{1}{a} + \frac{1}{b})}}\right), \tag{3.26}$$

and

$$P_m = 1 - Q\left(\frac{\lambda - \theta}{\sqrt{K(\frac{1}{a} + \frac{1}{b}) - \frac{4\theta}{b}}}\right), \tag{3.27}$$

respectively. Eliminating λ by combining (3.26) and (3.27), we can find P_f for given a, b, and P_m as follows,

$$P_f = Q\left(\frac{Q^{-1}(1 - P_m)\sqrt{K(\frac{1}{a} + \frac{1}{b}) - \frac{4\theta}{b}} + \theta}{K(\frac{1}{a} + \frac{1}{b})}\right), \tag{3.28}$$

where $Q^{-1}(\cdot)$ represents the inverse Q-function.

3.5 Numerical Results

In this section, some numerical and *monte carlo* simulation results are presented to demonstrate the performance of the proposed MF and EDC detectors for the proactive spectrum sensing.

3.5.1 MF Detector

Since we are interested in the ratio between f_m and f_x, the small values of f_m and f_x are used for facilitating simulations, i.e., the maximum doppler frequency is $f_m = 2\,\text{Hz}$. Different from existing spectrum sensing methods working in a low

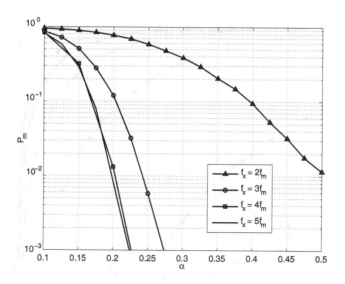

Fig. 3.4 Probability of missed detection in the MF detector

SNR region [8], our proactive approach works at the similar SNR as the nearby PR who are communicating with the PT under medium or even high SNR. Thus we set the SNR as 10 dB in our simulations. In addition, we assume that the period of energy detection samples, T, is 0.02 seconds and 10^4 *monte carlo* trails have been done for each curve. Figure 3.4 shows the probability of missed detection, P_m, in different scenarios, where the thresholds are chosen so as to satisfy $P_f \approx 10^{-2}$. From the figure, P_m goes down dramatically as interference strength, α, increases. In particular, $P_m < 10^{-2}$ when $\alpha > 0.18$ in $f_x = 5f_m$ case. Furthermore, Fig. 3.4 indicates that P_m goes down with the f_x-to-f_m ratio. This is because in the correlation operation at the MF detector, there will be spectrum overlaps between $\tilde{H}(f)$ and $\tilde{H}(f) \otimes \tilde{X}(f)$ when $f_x < 4f_m$, as shown in Fig. 3.2, which degrades the performance of the MF detector.

3.5.2 EDC Detector

Figure 3.5 demonstrates the ROC performance form (3.26) and (3.27) versus different SNRs at a CR user, γ_{CR}, where $\theta = 5$ dB. From the figure, the ROC performance improves as γ_{CR} goes up, i.e., the higher γ_{CR} is, the more ROC benefits can be obtained. Based on (3.28), Fig. 3.6 shows the probability of false alarm, P_f, versus the number of blocks in "silent" status of the pulse jamming signal, b. Here 30 blocks in "on" status of the pulse jamming signal are considered, i.e., $a = 30$, and both a and b determine the duty circle, $v = \frac{a}{a+b}$. We further assume that the number of samples during each observation block is $M = 10$, the SNR is $\gamma_{CR} = 10$ dB,

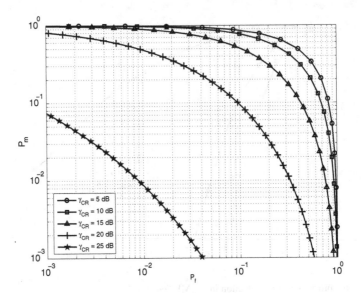

Fig. 3.5 ROC performance versus SNRs

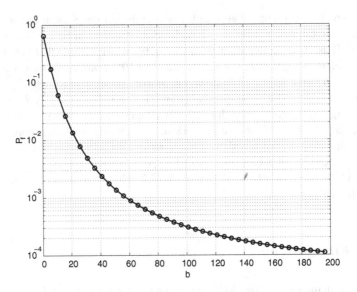

Fig. 3.6 False alarm probability versus the silent pulses duration in the EDC detector

the power adjustment at the PT is $\theta = 10\,\text{dB}$, and the threshold is selected so that $P_m \approx 10^{-2}$. It is shown in the figure that P_f decreases as b increases, in particular, $P_f < 10^{-2}$ when $b > 30$.

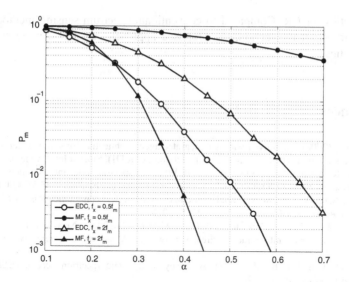

Fig. 3.7 Comparison of the MF and the EDC detectors

3.5.3 Comparison Between the MF and the EDC Detectors

Figure 3.7 compares the performance of the MF and the EDC detectors in different scenarios. In our simulation, $f_m = 4\,\text{Hz}$ and the thresholds are chosen to satisfy $P_f \approx 10^{-2}$ as well. In the EDC method, the duty cycle is $\nu = 0.5$ and the power of jamming pulses equals the average power of the sinusoid jamming signal in the MF detector. Therefore, the interference to the PR caused by the two proactive sensing algorithms is identical, but the EDC detector needs twice the duration of the MF detector to identify a PR. As shown in Fig. 3.7, the performance of the EDC detector in $f_x = 0.5f_m$ case is much better than that of the MF detector. However, as f_x increases from $f_x = 0.5f_m$ to $f_x = 2f_m$, the probability of missed detection in EDC, P_m, goes up while that of the MF detector goes down dramatically. This is because in fast fading channels, the sensing duration is relatively large so that the variation of the channel gains can be averaged and the power of PT's signal can be measured with a small variance, while in slow fading channels, the variation of the channel gains in the sensing duration will lead to a large variance of the received signal, which confuses the CR users and degrades the detection performance considerably.

3.6 Summary

In this chapter, a jamming-based probing method was proposed to detect primary receivers, which enables a CR system to coexist with a primary system without harmful interference to primary receivers. In particular, we developed two practical detectors and the corresponding jamming signals to identify the spectrum opportunities for CR. Compared to conventional spectrum sensing methods with primary transmitter detection, the proposed approach can obtain more spectrum opportunities.

References

1. Mitola, J. (2000). Cognitive radio: An integrated agent architecture for software defined radio. Ph.D. dissertation. The Royal Institute of Technology (KTH), Stockholm, Swedem.
2. Tandra, R., Mishra, M., & Sahai, A. (2009). What is a spectrum hole and what does it take to recognize one? *Proceedings of IEEE: Special Issues on Cognitive Radio, 97*(5), 824–848.
3. Haykin, S. (2005). Cognitive radio: Brain-empowered wireless communications. *IEEE Journal on Selected Areas in Communications, 23*(2), 201–220.
4. Akyildiz, I. A., Lee, W. Y., Vuran, M. C., & Mohanty, S. (2006). NeXt generation/dynamic spectrum access/cognitive radio wireless networks: A survey. *Computer Networks, 50,* 2127–2159.
5. Zhao, Q. & Sadler, B. M. (2007). A survey of dynamic spectrum access. *IEEE Signal Processing Magazine, 24,* 79–89.
6. Yucek, T. & Arslan, H. (2009). A survey of spectrum sensing algorithms for cognitive radio applications. *IEEE Communications Surveys and Tutorials, 11*(1), 116–130.
7. Digham, F. F., Alouini, M. S., & Simon, M. K. (2007). On the energy detection of unknown signals over fading channels. *IEEE Transactions on Communications, 55*(1), 21–24.
8. Tandra, R. & Sahai, A. (2008). SNR walls for signal detection. *IEEE Journal on Selected Topics in Singal Proccessing, 2*(1), 4–17.
9. Dandawate, A. V. & Giannakis, G. B. (1994). Statistical tests for presence of cyclostationarity. *IEEE Transactions on Signal Proccessing, 42*(9), 2355–2369.
10. Mishra, S. M., Sahai, A., & Brodersen, R. W. (2006). Cooperative sensing among cognitive radios. In *Proceedings of IEEE International Conference on Communications (ICC 2006),* Istanbul, pp. 1658–1663, 11–15 June 2006.
11. Ghasemi, A. & Sousa, E. S. (2005). Collaborative spectrum sensing for oppotunistic access in fading environment. In *Proceedings of IEEE International Symposium on New Frontiers in Dynamic Spectrum Access Networks (DySPAN 2005),* Baltimore, pp. 131–136, 8–11 November 2005.
12. Ghasemi, A. & Sousa, E. S. (2007). Opportunistic spectrum access in fading channels through collobrative sensing, *IEEE Journal on Communications, 2*(2), 71–82.
13. Hillenbrand, J., Weiss, T., & Jondral, F. K. (2005). Calculation of detection and false alarm probabilities in spectrum pooling systems. *IEEE Communications Leters, 9*(4), 349–351.
14. Gandetto, M. & Regazzoni, C. (2007). Spectrum sensing: A distributed approach for cognitive terminals. *IEEE Journal on Selected Areas in Communications, 25*(3), 546–557.
15. Ma, J., Zhao, G., & Li, Y. (2008). Soft combination and detection for cooperative spectrum sensing in cognitive radio networks. *IEEE Transactions on Wireless Communications, 7*(11), 4502–4507.
16. Ganesan, G. & Li, Y. (2007). Cooperative spectrum sensing in cognitive radio - part I: Two user networks. *IEEE Transactions on Wireless Communications, 6*(6), 2204–2213.
17. Ganesan, G. & Li, Y. (2007). Cooperative spectrum sensing in cognitive radio - part II: Multiuser networks. *IEEE Transactions on Wireless Communications, 6*(6), 2214–2222.
18. Goldsmith, A., Jafar, S. A., Maric, I., & Srinivasa, S. (2009). Breaking spectrum gridlock with cognitive radios: An information theoretic perspective. *Proceedings of IEEE: Special Issues on Cognitive Radio, 97*(5), 894–914.

19. Chakravarthy, V., Wu, Z., Temple, M., Garber, F., & Li, X. (2008). Cognitive radio centric overlay/underlay waveform. In *Proceedings of IEEE International Symposium on New Frontiers in Dynamic Spectrum Access Networks (DySPAN 2008)*, Chicago, IL, pp. 1–10, 14–17 October 2008.
20. Sahin, M. E., Ahmed, S., & Arslan, H. (2007). The roles of ultra wideband in cognitive networks. In *Proceedings of IEEE International Conference on Ultra-Wideband (ICUWB 2007)*, Singapore, pp. 247–252, 24–26 September 2007.
21. Mishra, S. M. & Brodersen, R. W. (2007). Cognitive technology for improving ultra-wideband (UWB) coexistence. In *Proceedings of IEEE International Conference on Ultra-Wideband (ICUWB 2007)*, Singapore, pp. 253–258, 24–26 September 2007.
22. Wild, B. & Ramchandran, K. (2005). Detecting primary receivers for cognitive radio applications. In *Proceedings of IEEE International Symposium on New Frontiers in Dynamic Spectrum Access Networks (DySPAN 2005)*, Baltimore, pp. 124–130, 8–11 November 2005.
23. Stuber, G. L. (2002). *Principles of mobile communication (2nd ed.)*. Berlin: Springer.
24. Huang, Y. C. & Polydoros, A. (1995). Likelihood methods for mpsk modulation classification. *IEEE Transactions on Communications, 43*(2/3/4), 1493–1504.
25. Schreyogg, C. (1997). Modulation classification of QAM schemes using the DFT of phase histogramm combined with modulus information. In *Proceedings of IEEE Military Communications Conference (MILCOM 1997)*, Monterey, CA, pp. 1372–1376, 2–5 November 1997.
26. Urkowitz, H., (1967). Energy detection of unknown deterministic signals. *Proceedings of IEEE, 55*(4), 523–531.
27. Rappaport, T. S. (2001). *Wireless communications: principles and practice*. Upper Saddle River: Prentice Hall.

Chapter 4
Relay-Based Probing for Spectrum Sensing

4.1 Introduction

In spectrum sharing, the cross-channel gain from the *cognitive transmitter* (CT) to the *primary receiver* (PR) significantly affects the cognitive capacity [1, 2]. When the cross-channel gain is available, the CT can precisely control its interference to the PR and achieve significant cognitive capacity. However, when the cross-channel gain is not available, the CT has to reduce the transmission power to protect the PR, which inevitably degrades the cognitive capacity. Therefore, efficient spectrum sharing schemes require the cross-channel gain [3].

In practice, estimating the cross-channel gain is a very challenging task. In *frequency-division duplex* (FDD) systems, the cross-channel gain can only be estimated by the PR and sent back to the CT via the backhaul link between the two systems. However, such a backhaul assumption is usually invalid in cognitive radio networks [4]. In *time-division duplex* (TDD) systems, the cross-channel gain can be obtained by the CT in the PR's reverse transmission if the CT knows the transmission power of the PR. However, such an assumption is also invalid [4]. Therefore, the conventional estimation methods [5–7] are not suitable for cognitive radio systems and estimating the cross-channel gain becomes the bottleneck for efficient spectrum sharing.

Recently, a new category of estimation methods, called proactive estimation, is proposed in [8–14] to estimate the cross-channel gain, which does not need either the backhaul link or the transmission power of the PR. In proactive estimation, the CT transmits some jamming signals to probe the *close-loop power control* (CLPC) between the *primary transmitter* (PT) and PR, i.e., the jamming signals pass through the cross-channel, degrade the *signal-to-noise ratio* (SNR) at the PR, and force the

© The Author(s) 2017
G. Zhao et al., *Advanced Sensing Techniques for Cognitive Radio*,
SpringerBriefs in Electrical and Computer Engineering,
DOI 10.1007/978-3-319-42784-3_4

PT to adjust the transmission power to maintain a certain target SNR at the PR. Then the power adjustment of the PT becomes a function of the cross-channel gain. As a result, by measuring the power adjustment of the primary signal, the CT is able to autonomously estimate the cross-channel gain. In [8], the proactive estimation is first proposed to obtain the cross-channel gain in cognitive radio networks. In [10, 11], the power of the jamming signals is designed to optimize the estimation performance. In [12–14], the proactive estimation is further developed in multiple antenna systems to obtain the null space or the channel direction information of the cross-channel.

However, all these proactive sensing methods require the CT to transmit jamming signals, which may inevitably cause severe interference to the PR. This raises a new issue in the sensing phase, called *spectrum sensing interference*. Conventionally, spectrum sensing does not cause any interference to the PR since it works passively. Therefore, we need to strictly control the sensing interference to an extremely low level when designing the proactive sensing algorithms. Otherwise, it is difficult to implement the proactive sensing in practical systems.

In this chapter, the motivation is to deal with the spectrum sensing interference caused by the jamming-based proactive estimation methods. A relay-based method is proposed to conduct the proactive estimation. In our method, the CT acts as a full-duplex *amplify-and-forward* (AF) relay. This allows the CT to use the relayed primary signal rather than the jamming signal for the probing. As a result, the interference to the PR can be effectively reduced.

4.2 System Model

Figure 4.1 provides the system model of this chapter, where the PT serves the PR that is uniformly located inside the disk region with the center PT and the radius R. At the same time, a CT intends to estimate the cross-channel gain from the CT to the PR for spectrum sharing, and the distance between the CT and the PT is r. Here, we denote $h_k \sqrt{g_k}$ ($k = 0, 1,$ and 2) as the channel coefficients among the three nodes, where h_k and g_k represent the small-scale fading and large-scale path loss coefficients, respectively. In the small-scale fading, the coefficient h_k follows Rayleigh distribution with unit variance. In the large-scale path loss [15], the coefficient g_k follows

$$g_k(\text{dB}) = -128.1 - 37.6 \log_{10}(l) \quad \text{for} \quad l \geq 0.035 \text{ km}, \tag{4.1}$$

where l denotes the distance between two nodes and the system is assumed to operate over 2 GHz frequency band.

Furthermore, the block fading channel is considered, where the Rayleigh fading coefficients are constant within each block and they are independent for different blocks. It is also assumed that all nodes are stationary and their path loss coefficients are constant in different blocks.

Fig. 4.1 System model

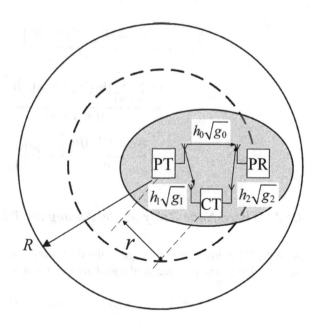

In the following, the point-to-point model and the three-node relay model are introduced, respectively.

4.2.1 Point-to-Point Model Between the PT and PR

Denote $x(i,j)$ as the transmitted signal of the PT with unit power, i.e., $\mathbb{E}\left[|x(i,j)|^2\right] = 1$, where $\mathbb{E}[\cdot]$ is the expectation operator, and i and j denote the indices of N samples and M blocks. Then the received signal at the PR can be expressed as

$$y(i,j) = h_0(j)\sqrt{g_0 p_0}x(i,j) + n_p(i,j), \qquad (4.2)$$

where p_0 is the transmission power of the PT and $n_p(i,j)$ is the *addictive white Gaussian noise* (AWGN) at the PR with zero mean and variance σ^2.

The guaranteed primary service with CLPC is assumed, where the PT automatically adjusts its transmission power to maintain a certain target average SNR or *signal-to-interference-plus-noise ratio* (SINR), denoted as $\bar{\gamma}_T$. In particular, the value $\bar{\gamma}_T$ is publicly available knowledge and it is known to the CT. The relationship between $\bar{\gamma}_T$ and p_0 can be obtained by

$$\bar{\gamma}_T = \frac{\mathbb{E}\left[\left|h_0(j)\sqrt{g_0P_0}x(i,j)\right|^2\right]}{\sigma^2}$$

$$= \frac{\sum\limits_{j=1}^{M}\sum\limits_{i=1}^{N}\left|h_0(j)\sqrt{g_0P_0}x(i,j)\right|^2}{M \times N \times \sigma^2} \tag{4.3}$$

$$= \frac{P_0g_0\mathbb{E}\left[\left|h_0(j)\right|^2\right]}{\sigma^2} = \frac{P_0g_0}{\sigma^2}.$$

4.2.2   Three-Node Relay Model Among the PT, CT, and PR

Since the CT is inside the coverage of the PT for spectrum sharing, it can overhear the PT's signal. Then the received signal at the CT can be expressed as

$$x_{cr}(i,j) = h_1(j)\sqrt{g_1P_0}x(i,j) + n_c(i,j), \tag{4.4}$$

where $n_c(i,j)$ is the AWGN at the CT with zero mean and the variance σ^2. The corresponding average SNR at the CT can be obtained by

$$\bar{\gamma}_{c_0} = \frac{\mathbb{E}\left[\left|h_1(j)\sqrt{g_1P_0}x(i,j)\right|^2\right]}{\sigma^2}$$

$$= \frac{P_0g_1\mathbb{E}\left[\left|h_1(j)\right|^2\right]}{\sigma^2} = \frac{P_0g_1}{\sigma^2}. \tag{4.5}$$

When the CT acts as a full-duplex AF relay with the amplitude gain G, the transmitted signal of the CT becomes

$$x_{ct}(i,j) = Gx_{cr}(i,j). \tag{4.6}$$

In fact, there is a time delay in full-duplex relay between reception and transmission, called *signal processing delay*. However, the impact of the signal processing delay depends on the relationship between the signal processing delay and the system sample period. If the signal processing delay is relative small compared with the system sample period, the destination cannot distinguish the direct and relay paths. Therefore, the signal processing delay can be ignored in the baseband signal processing model in (4.6). This is in particular valid in AF full-duplex relay systems since the AF relay can be implemented through RF circuits with extremely short signal processing delay. This assumption has also been widely used in existing literature on full-duplex relay systems, e.g., [16–18].

Since this chapter focuses on proposing a new sensing technique, it does not consider the specific self-interference suppression techniques. Instead, it is assumed that most of the self-interference can be cancelled using existing self-interference suppression methods. The residual self-interference (after self-interference suppression) is modelled as the noise. Then the received signal of the PR has two components: the direct signal from the PT and the relay signal from the CT. In addition, since the two components arrive at the PR via different paths, τ is defined as the *time-difference-of-arrival* (TDOA) [19] between them. Then the overall received primary signal at the PR can be expressed as

$$y_p = \underbrace{h_0(j)\sqrt{g_0 p_0}x(i,j)}_{S_d(\text{Direct signal})}$$

$$+ \underbrace{Gh_2(j)\sqrt{g_2}h_1(j)\sqrt{g_1 p_0}x(i-\alpha, j-\beta)}_{S_r(\text{Relay signal})}$$

$$+ \underbrace{\rho Gh_2(j)\sqrt{g_2}n_c(i,j) + n_p(i,j)}_{N(\text{Noise})}, \tag{4.7}$$

where ρ is the parameter indicating the strength of the residual self-interference in full-duplex relay, α and β are delay indices. As a result, the TDOA becomes $\tau = \alpha T_s + \beta T_b$, where T_s and T_b represent the block period and sample period, respectively.

4.3 Estimation Principle

The motivation of this chapter is to deal with the spectrum sensing interference caused by proactive sensing. In conventional proactive sensing, the jamming signal is used to probe the CLPC between the primary transceivers, which allows the CT to autonomously estimate the cross-channel gain. However, as the side effect, using jamming signal inevitably causes the sensing interference to the PR. In particular, the more sensing interference the CT generates, the more effectively the CLPC can be probed, and then the better the estimation performance becomes. To deal with the sensing interference, the proposed solution is to let the CT act as a full-duplex AF relay to conduct the probing, i.e., replace the jamming signal by the primary signal using the full-duplex AF relay. Therefore, the PR can receive the desired signals from both PT and CT, which can effectively reduce the sensing interference to the PR.

In principle, when the CT conducts the full-duplex AF relay, it actually changes the original point-to-point channel between the primary transceivers to the three-node relay channel. Then the original PT-PR channel gain g_0 becomes the *equivalent end-to-end channel gain* (EEECG) g_e. If $g_0 \neq g_e$, the received SNR of the PR becomes unequal to the target SNR. Then the CLPC can be triggered to adjust the transmission power of the PT. As a result, by observing the power adjustment, the CT can estimate the cross-channel gain.

Depending on the relationship between g_0 and g_e, the CLPC is triggered in different ways, which has different impacts on the primary link. If $g_0 < g_e$, it means that the CT enhances the primary link. Then the SNR of the PR becomes larger than the target value, and the PT reduces the power to maintain the target SNR. On the other hand, if $g_0 > g_e$, it means that the CT causes interference and degrades the primary link. Then, the SNR of the PR becomes less than the target value, and the PT increases the power to maintain the target SNR. Therefore, to avoid interfering with the PR, the relay-based probing needs to work in the case of $g_0 < g_e$ since the case of $g_0 > g_e$ causes the interference to the PR.

In the rest of this section, the EEECG expressions are obtained and then the relationship between g_0 and g_e is analyzed to find two interference-free regions, in which the CT's probing can render $g_0 < g_e$. In the next section, the new algorithm is proposed to identify the CT located region and estimate the cross-channel gain.

4.3.1   Calculate the EEECG g_e

As indicated before, since the transmitted signal of the PT travels through different paths, the PR may receive multiple copies of the signal, which arrive at the PR at different time. Depending on the system bandwidth and also the signal processing ability, the PR may only collect the received signals within a ceratin time duration, which is called the maximum allowable TDOA [19] and defined as T_m. If the TDOA of the direct and relay signals is less than T_m, i.e., $\tau < T_m$, it is the *small delay case*, where the PR treats both of them as the desired signals. Otherwise, if the TDOA is equal to or lager than T_m, i.e., $\tau \geq T_m$, it is the *large delay case*, where the PR only treats one of them as the desired signal and leaves the other as the interference. In the following, the EEECG expressions are developed under the small and large delay cases, respectively.

Small Delay Case: When $\tau < T_m$, the PR treats both direct and relay signals as the desired signals. Then the average SNR at the PR can be obtained by

$$
\begin{aligned}
\bar{\gamma} &= \mathbb{E}\left[\frac{|S_d + S_r|^2}{|N|^2}\right] \\
&= \mathbb{E}\left[\frac{\left|h_0(j)\sqrt{g_0 p_0}x(i,j)+Gh_1(j)h_2(j)\sqrt{g_1 g_2 p_0}x(i-\alpha,j-\beta)\right|^2}{\left(\left|\rho h_2(j)G\sqrt{g_2}\right|^2+1\right)\sigma^2}\right] \\
&\approx \frac{p_0 g_0 + G^2 p_0 g_1 g_2}{(\rho^2 G^2 g_2 + 1)\sigma^2} = \frac{p_0 g_e}{\sigma^2},
\end{aligned}
\tag{4.8}
$$

where the EEECG is

$$g_e = \frac{g_0 + G^2 g_1 g_2}{\rho^2 G^2 g_2 + 1}.$$

(4.9)

Large Delay Case: When $\tau \geq T_m$, the PR automatically treats the strong one (between the direct and relay signals) as the desired signal and leaves the other as the interference. Since either the direct signal from the PT or the relay signal from the CT can be the strong one, we discuss them in the following two subcases.

- *Strong Direct Signal*: The PR treats the direct signal as its desired signal as long as the direct signal can provide higher SINR than that the relay signal can provide, i.e.,

$$\mathbb{E}\left[\frac{|S_d|^2}{|S_r|^2 + |N|^2}\right] > \mathbb{E}\left[\frac{|S_r|^2}{|S_d|^2 + |N|^2}\right].$$

(4.10)

Then the average SINR at the PR can be obtained by

$$\bar{\gamma}' = \mathbb{E}\left[\frac{|S_d|^2}{|S_r|^2 + |N|^2}\right]$$

$$\approx \frac{1}{2}\left(\frac{p_0 g_0}{G^2 p_0 g_1 g_2 + \rho^2 G^2 g_2 \sigma^2 + \sigma^2} + \frac{p_0 g_0}{\sigma^2}\right) = \frac{p_0 g_e}{\sigma^2},$$

(4.11)

where the EEECG becomes

$$g_e = \frac{g_0}{2}\left(\frac{1}{\frac{G^2 p_0 g_1 g_2}{\sigma^2} + \rho^2 G^2 g_2 + 1} + 1\right).$$

(4.12)

- Strong Relay Signal: The PR treats the relay signal as its desired signal as long as the relay signal can provide equal or higher SINR than that the direct signal can provide, i.e.,

$$\mathbb{E}\left[\frac{|S_d|^2}{|S_r|^2 + |N|^2}\right] \leq \mathbb{E}\left[\frac{|S_r|^2}{|S_d|^2 + |N|^2}\right].$$

(4.13)

Then the average SINR at the PR can be obtained by

$$\bar{\gamma}'' = \mathbb{E}\left[\frac{|S_r|^2}{|S_d|^2 + |N|^2}\right] \approx \frac{G^2 p_0 g_1 g_2}{p_0 g_0 + \rho^2 G^2 g_2 \sigma^2 + \sigma^2} = \frac{p_0 g_e}{\sigma^2},$$

(4.14)

where the EEECG becomes

$$g_e = \frac{G^2 g_1 g_2}{\frac{p_0 g_0}{\sigma^2} + \rho^2 G^2 g_2 + 1}.$$

(4.15)

Once we obtain the SNRs in (4.11) and (4.14), we can substitute them into (4.10) and have

$$\frac{1}{2}\Big(\frac{p_0 g_0}{G^2 p_0 g_1 g_2 + \rho^2 G^2 g_2 \sigma^2 + \sigma^2} + \frac{p_0 g_0}{\sigma^2}\Big) > \frac{G^2 p_0 g_1 g_2}{p_0 g_0 + \rho^2 G^2 g_2 \sigma^2 + \sigma^2}. \qquad (4.16)$$

When we further substitute (4.3) and (4.5) into (4.16) and obtain

$$\begin{aligned}
&\bar{\gamma}_T(\bar{\gamma}_T + \rho^2 G^2 g_2 + 1) + \bar{\gamma}_T(G^2 \bar{\gamma}_{c_0} g_2 + \rho^2 G^2 g_2 + 1)(\bar{\gamma}_T + \rho^2 G^2 g_2 + 1) \\
&> 2G^2 \bar{\gamma}_{c_0} g_2 (G^2 \bar{\gamma}_{c_0} g_2 + \rho^2 G^2 g_2 + 1),
\end{aligned} \qquad (4.17)$$

we can have the following inequality

$$G < \sqrt{\frac{a}{g_2}}, \qquad (4.18)$$

where

$$\begin{aligned}
a = &\frac{-(\bar{\gamma}_{c_0}\bar{\gamma}_T\bar{\gamma}_T + \rho^2\bar{\gamma}_T\bar{\gamma}_T + \bar{\gamma}_{c_0}\bar{\gamma}_T + 3\rho^2\bar{\gamma}_T - 2\bar{\gamma}_{c_0})}{2(\bar{\gamma}_{c_0} + \rho^2)(\rho^2\bar{\gamma}_T - 2\bar{\gamma}_{c_0})} \\
&- \frac{\sqrt{(\bar{\gamma}_{c_0}\bar{\gamma}_T\bar{\gamma}_T + \rho^2\bar{\gamma}_T\bar{\gamma}_T + \bar{\gamma}_{c_0}\bar{\gamma}_T + 3\rho^2\bar{\gamma}_T - 2\bar{\gamma}_{c_0})^2 - 4(\bar{\gamma}_{c_0} + \rho^2)(\rho^2\bar{\gamma}_T - 2\bar{\gamma}_{c_0})(2\bar{\gamma}_T\bar{\gamma}_T + 2\bar{\gamma}_T)}}{2(\bar{\gamma}_{c_0} + \rho^2)(\rho^2\bar{\gamma}_T - 2\bar{\gamma}_{c_0})}.
\end{aligned} \qquad (4.19)$$

This indicates that when the amplitude gain G is less than $\sqrt{a/g_2}$, the PR treats the direct signal as the desired signal. Otherwise, the PR treats the relay signal as the desired signal.

4.3.2 Relationship Between g_0 and g_e

Based on the above EEECG expressions, this subsection discusses the relationship between the original PT-PR channel gain g_0 before the relay and the EEECG g_e after the relay. The goal is to find the interference-free regions, in which the CT can conduct the relay-based probing without interfering with the PR, i.e., let $g_0 < g_e$ always hold. In the following, two theorems are presented to show the relationship.

Theorem 1. *In the small delay case, the relay enhances (or degrades) the primary link if the first hop of the relay channel is stronger (or weaker) than the primary channel, i.e.,*

$$\begin{cases} g_0 < g_e, & \text{if } g_1 > \rho^2 g_0, & (4.20a) \\ g_0 \geq g_e, & \text{if } g_1 \leq \rho^2 g_0. & (4.20b) \end{cases}$$

Proof. The detailed proof can be found in [20].

Theorem 2. *In the large delay case, the relay enhances the primary link if the first hop of the relay channel is stronger than the primary channel and the amplitude gain is greater than the value $\sqrt{b/g_2}$. Otherwise, the relay degrades the primary link if the first hop of the relay channel is stronger than the primary channel but the amplitude gain of the relay is no more than $\sqrt{b/g_2}$, or if the first hop of the relay channel is not stronger than the primary channel, i.e.,*

$$
\begin{cases}
g_0 < g_e, & \text{if } g_1 > \rho^2 g_0 \text{ and } G > \sqrt{\frac{b}{g_2}}, & (4.21a) \\[2ex]
g_0 \geq g_e, & \text{if } g_1 \leq \rho^2 g_0 \text{ or } G \leq \sqrt{\frac{b}{g_2}}. & (4.21b)
\end{cases}
$$

Proof. The detailed proof can be found in [20].

The above two theorems indicate that the relationship between g_0 and g_e is mainly determined by three factors:

1) in which case the relay is operating, the small or large delay case?
2) what is the relationship between g_0 and g_1, $g_1 > \rho^2 g_0$ or $g_1 \leq \rho^2 g_0$?
3) whether or not the amplitude gain $G > \sqrt{b/g_2}$ is satisfied?

- *The First Two Factors:* Since the first two factors are determined by the location of the CT, Fig. 4.2 is provided to show the relationship. In the figure, the ellipse and the dashed circle divide the whole coverage of the PT into four regions, called Regions I, II, III, and IV. For the ellipse, the PT and PR are located at the two focuses, and the boundary of the ellipse is determined by the maximum allowable TDOA T_m. If the CT is located at any point of the ellipse, the TDOA of the direct and relay signals is equal to T_m. Thus, the regions inside the ellipse (Regions I

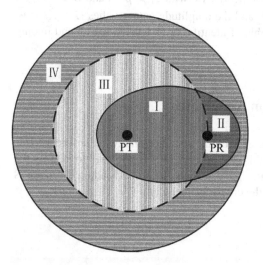

	$\tau < T_m$	$\tau \geq T_m$
$g_1 > \rho^2 g_0$	I	III
$g_1 \leq \rho^2 g_0$	II	IV

Fig. 4.2 Four location regions for the CT

and II) and those outside the ellipse (Regions III and IV) are corresponding to the small and large delay cases, respectively. For the circle, the PT is located at the center and the radius is the PT-PR distance. Therefore, the regions inside the circle (Regions I and III) and those outside the circle (Regions II and IV) are corresponding to $g_1 > \rho^2 g_0$ and $g_1 \leq \rho^2 g_0$, respectively.

In practice, since the CT may appear in any one of the four regions, it needs to identify its located region. This is because the CT can conduct the relay-based probing without interfering with the PR only in Regions I and III, i.e., the inequality $g_0 < g_e$ only holds for $g_1 > \rho^2 g_0$.

- *The Third Factor*: Since the relationship between g_0 and g_e is also affected by the third factor, i.e., the amplitude gain G, Fig. 4.3 is provided to show how the amplitude gain affects the EEECG g_e, where the same system parameters are adopted as in Sect. 4.5. Here, a specific CT location in each region is considered and four EEECG curves are provided based on (4.9), (4.12), and (4.15). Figure 4.3a considers Regions I and III where the first hop of the relay channel is stronger than the primary channel, i.e., $g_1 > \rho^2 g_0$. For the CT in Region I, the EEECG is always greater than the primary channel gain, i.e., $g_e > g_0$, which agrees with Theorem 1. For the CT in Region III, the EEECG is less (or greater) than the primary channel gain if the amplitude gain of the relay is less (or greater) than the value $\sqrt{b/g_2}$, which agrees with Theorem 2. Figure 4.3b considers Regions II and IV where the primary channel is stronger than or equal to the first hop of the relay channel, i.e, $g_1 \leq \rho^2 g_0$. From the figure, both EEECG curves are always less than or equal to the primary channel gain, i.e., $g_e \leq g_0$. This is true and agrees with the two theorems.

From the above analysis, the relay-based probing causes no interference to the PR only in two case: one is the small delay case with $g_1 > \rho^2 g_0$, i.e., the CT is located in Region I, and the other is the large delay case with $g_1 > \rho^2 g_0$ and $G > \sqrt{b/g_2}$, i.e., the CT is located in Region III and the amplitude gain satisfies $G > \sqrt{b/g_2}$. Therefore, the CT needs to be capable of identifying its located region and finding the value $\sqrt{b/g_2}$.

4.4   Probing and Estimation Algorithms

In the previous section, the relay-based probing is studied and two interference-free regions are obtained for the CT. In this section, a method is proposed to identify the CT located region and find the value $\sqrt{b/g_2}$. Then, the corresponding estimator is developed in each region to obtain the cross-channel gain.

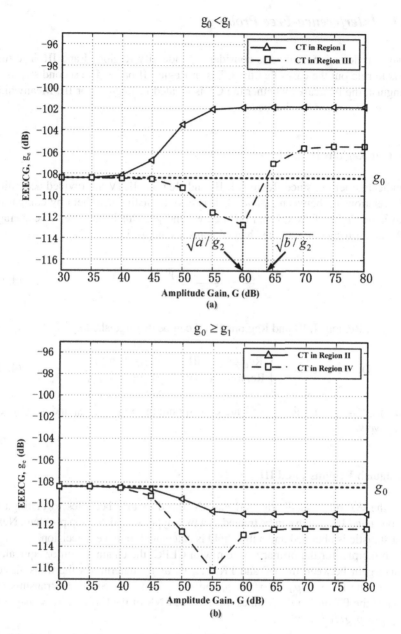

Fig. 4.3 An example to demonstrate the relationship between g_0 and g_e

4.4.1 Interference-Free Probing

A two-step detection method is considered to identify Regions I and III. The first step is to rule out the cases that the CT is in Region II or IV. The second step is to distinguish the two cases whether the CT is located in Region I or III. Meanwhile, the value $\sqrt{b/g_2}$ can be obtained.

Rule Out Regions II and IV

As indicated before, since Regions I, III and Regions II, IV are divided according to the relationship between g_0 and g_1, it is easy to distinguish them by comparing the average SNRs of the CT and PR, i.e., γ_{c_0} and γ_T. This is because these average SNRs are corresponding to the channel gains g_0 and g_1, i.e.,

$$\frac{g_1}{g_0} = \frac{\frac{\bar{\gamma}_{c_0}\sigma^2}{p_0}}{\frac{\bar{\gamma}_T\sigma^2}{p_0}} = \frac{\bar{\gamma}_{c_0}}{\bar{\gamma}_T}. \tag{4.22}$$

Therefore, Regions I, III and Regions II, IV can be distinguished by

$$\text{Decision result} = \begin{cases} \text{Region I or III,} & \text{if } \bar{\gamma}_{c_0} > \rho^2\bar{\gamma}_T, \\ \text{Region II or IV,} & \text{if } \bar{\gamma}_{c_0} \leq \rho^2\bar{\gamma}_T, \end{cases} \tag{4.23}$$

where $\bar{\gamma}_{c_0} > \rho^2\bar{\gamma}_T$ and $\bar{\gamma}_{c_0} \leq \rho^2\bar{\gamma}_T$ are corresponding to $g_1 > \rho^2 g_0$ and $g_1 \leq \rho^2 g_0$, respectively.

Distinguish Regions I and III

Once the CT rules out Regions II and IV, the difficulty becomes to distinguish Regions I and III, which can be treated as a two-hypothesis detection problem. Next, a test statistic is obtained and a threshold is calculated to make a decision.

In principle, for the primary system with CLPC, the channel gain or equivalent channel gain between the PT and PR, i.e., g_0 or g_e, determines the transmission power of the PT, i.e., $p_0 = \sigma^2\bar{\gamma}_T/g_0$ and $p_1 = \sigma^2\bar{\gamma}_T/g_e$. Since the transmission power of the PT further determines the average SNR of the CT, i.e., $\bar{\gamma}_{c_0} = p_0 g_1/\sigma^2$ and $\bar{\gamma}_{c_1} = p_1 g_1/\sigma^2$, it has

$$\frac{\bar{\gamma}_{c_0}}{\bar{\gamma}_{c_1}} = \frac{p_0}{p_1}, \tag{4.24}$$

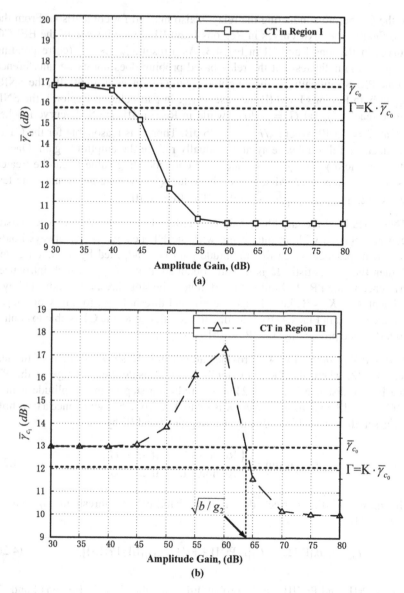

Fig. 4.4 The principle of obtaining the test statistic Ω in distinguish Regions I and III

which indicates that the relationship between g_0 and g_e is corresponding to the relationship of the measured SNRs at the CT before and after the relay, i.e., $\bar{\gamma}_{c_0}$ and $\bar{\gamma}_{c_1}$. Thus, it is possible for the CT to distinguish Regions I and III.

- *Find the Test Statistic*: Fig. 4.4 provides the measured SNR $\bar{\gamma}_{c_1}$ of the CT in Regions I and III versus the amplitude gain G, where the measured SNR $\bar{\gamma}_{c_0}$

of the CT before conducting the relay is also marked for comparison. From the two figures, the SNR curves in Regions I and III are the inverse of the EEECG curves in Regions I and III in Fig. 4.3. As a result, $\bar{\gamma}_{c_0} > \bar{\gamma}_{c_1}$ (corresponding to $g_0 < g_1$) indicates that the relay-based probing does not cause interference to the PR. Specifically, as the amplitude gain G reduces from 80 dB, the SNRs $\bar{\gamma}_{c_1}$ in both Regions I and III increase to $\bar{\gamma}_{c_0}$. In particular, in Region I, the SNR curve reaches $\bar{\gamma}_{c_0}$ at $G = \sqrt{b/g_2}$ while in Region III, the SNR curve reaches $\bar{\gamma}_{c_0}$ at $G = 35$ dB, where $\sqrt{b/g_2} \gg 35$ dB. Thus, it is reasonable for the CT to conduct the full-duplex relay and gradually reduce the amplitude gain. Once a stop threshold $\bar{\gamma}_{c_1} = \bar{\gamma}_{c_0}$ is satisfied, the corresponding value G can be treated as the test statistic Ω. Meanwhile, the value $\sqrt{b/g_2}$ is obtained since the test statistic Ω is equal to $\sqrt{b/g_2}$ for the CT in Region III.

Theoretically, if the amplitude gain G is reduced continuously, the test statistic Ω can be obtained without interfering with the PT, i.e., $\bar{\gamma}_{c_0} > \bar{\gamma}_{c_1}$ always holds. However, in practice, since the amplitude gain G is adjusted by a reducing step ΔG, then the test statistic Ω is actually obtained at $\bar{\gamma}_{c_0} < \bar{\gamma}_{c_1}$, which introduces interference to the PR. To limit the interference, the stop threshold is adjusted by a coefficient $0 < K \le 1$, i.e., change the original threshold γ_{c_0} to a new one $K\gamma_{c_0}$, which are shown in Fig. 4.4. By selecting the coefficient K, the CT is able to control the interference probability, i.e., $P_I = \Pr\{\bar{\gamma}_{c_0} < \bar{\gamma}_{c_1}\}$.

- *Find the Threshold*: Fig. 4.5 provides an example to show the PDF of the test statistic Ω, where the PR is randomly located inside the coverage of the PT, the PT-CT distance is $r = 0.2$ km, the reducing step of the amplitude gain is $\Delta G = 0.5$ dB, and the coefficient K is set to 0.9. From the figure, once a threshold G_t is set, the CT is able to distinguish Regions I and III by

$$\text{Decision result} = \begin{cases} \text{Region I,} & \text{if } \Omega < G_t, \\ \text{Region III,} & \text{if } \Omega \ge G_t. \end{cases} \tag{4.25}$$

To maximize the correct detection probability, the threshold G_t can be obtained by

$$P_d = \max_{G_t}[\Pr\{\Omega < G_t|\text{I}\}\Pr\{\text{I}\} + \Pr\{\Omega \ge G_t|\text{III}\}\Pr\{\text{III}\}], \tag{4.26}$$

where $\Pr\{\text{I}\}$ and $\Pr\{\text{III}\}$ are the probabilities that the CT is in Regions I and III, respectively.

4.4.2 Cross-Channel Gain Estimation

In this subsection, two estimators are designed for Regions I and III, respectively.

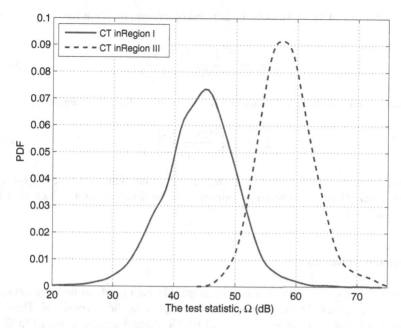

Fig. 4.5 The PDF of the test statistic Ω

Estimator in Region I

If the CT is located in Region I, it can conduct the full-duplex relay and change the original PT-PR channel gain g_0 to the EEECG g_e in (4.9). Then the received average SNR at the PR is changed to a new value. To maintain the target SNR $\bar{\gamma}_T$, the PT adjusts the transmission power to

$$p_1 = \frac{\bar{\gamma}_T \sigma^2}{g_e}. \tag{4.27}$$

Then, the measured average SNR at the CT becomes

$$\bar{\gamma}_{c_1} = \frac{p_1 g_1}{\sigma^2}. \tag{4.28}$$

From (4.5) and (4.28), it leads to the following relationship,

$$\frac{p_0}{p_1} = \frac{\bar{\gamma}_{c_0}}{\bar{\gamma}_{c_1}}. \tag{4.29}$$

By substituting (4.3), (4.5), (4.9), (4.28), and (4.29) into (4.27), it has

$$\bar{\gamma}_T = \frac{\frac{\bar{\gamma}_{c_1}}{\bar{\gamma}_{c_0}}\bar{\gamma}_T + G^2\bar{\gamma}_{c_1}g_2}{\rho^2 G^2 g_2 + 1}. \tag{4.30}$$

Therefore, the following estimator can be obtained for the CT in Region I, i.e.,

$$\hat{g}_2 = \frac{\bar{\gamma}_T(\bar{\gamma}_{c_1} - \bar{\gamma}_{c_0})}{G^2\bar{\gamma}_{c_0}(\rho^2\bar{\gamma}_T - \bar{\gamma}_{c_1})}. \tag{4.31}$$

In (4.31), since the relay-based probing causes no interference to the PR for any value of G, G can be chosen to minimize the estimation error, which will be discussed latter.

Estimator in Region III

If the CT is located in Region III, the amplitude gain needs to be large enough, i.e., $G > \sqrt{b/g_2}$, to let the relay signal be stronger than the direct signal. Then the relay-based probing changes the original PT-PR channel gain g_0 to the EEECG g_e in (4.15). Consequently, the SINR at the PR changes from $\bar{\gamma}_T$ to $\bar{\gamma}$ in (4.14). To maintain the target average SNR $\bar{\gamma}_T$, the PT adjusts the transmission power to a new value p_1, i.e.,

$$p_1 = \frac{\bar{\gamma}_T\sigma^2(\bar{\gamma}_T + \rho^2 G^2 g_2 + 1)}{G^2 g_1 g_2}. \tag{4.32}$$

At the CT, the measured average SNR becomes

$$\bar{\gamma}_{c_1} = \frac{p_1 g_1}{\sigma^2} = \frac{\bar{\gamma}_T(\bar{\gamma}_T + \rho^2 G^2 g_2 + 1)}{G^2 g_2}. \tag{4.33}$$

Therefore, the following estimator can be obtained for the CT in Region III, i.e.,

$$\hat{g}_2 = \frac{\bar{\gamma}_T^2 + \bar{\gamma}_T}{G^2(\bar{\gamma}_{c_1} - \rho^2\bar{\gamma}_T)}, \tag{4.34}$$

where $G > \sqrt{b/g_2}$. In (4.34), since the relay-based probing only causes no interference to the PR for $G > \sqrt{b/g_2}$, it needs to consider both the estimation error and the condition $G > \sqrt{b/g_2}$ when choosing the value of G, which will also be discussed latter.

4.5 Simulation Results

In this section, the simulation results are provided to demonstrate the performance of the proposed method. Here, the same system model is adopted as shown in Fig. 4.1, where the PT is located in the center of the disk with the radius $R = 0.5$ km, the PR is uniformly distributed on the disk, and the distance between the PT and CT is r km. In the simulation, the target average SNR of the PR is $\bar{\gamma}_T = 10$ dB, the reducing step of the amplitude gain is $\Delta G = 0.5$ dB, the maximum allowable TDOA at the PR is $T_m = 10^{-6}$ second, the noise power is -114 dBm, the number of blocks is $M = 200$, the number of samples in each block is $N = 200$, and the number of Monte Carlo trails is 10^3. For the wireless channels among the three nodes, the path loss, shadowing, and small-scale fading are considered, where the path loss coefficient is determined by the model in (4.1), the shadowing coefficient follows log-normal distribution with the standard deviation of 4, and the small-scale fading coefficient follows Rayleigh distribution with mean $\lambda_k = 1$ ($k = 0$, 1, and 2). When the impacts of the imperfect *self-interference suppression* (SIS) is considered, it has $\rho = 1.2598$, i.e., raise the noise floor at the CT by 2 dB according to [21, 22].

In the following, the coefficient K and the threshold G_t are determined to identify the CT located region and find the value $\sqrt{b/g_2}$. Then, the amplitude gain G is further determined to estimate the cross-channel gain. Finally, the performance of the proposed method is provided, which also compares with the conventional jamming-based method in [11].

4.5.1 Determine the Coefficient K and the Threshold G_t

Figure 4.6 provides the interference probability of the relay-based probing for different PT-CT distances. From the figure, as the value of K approaches 1, the interference probability grows. Furthermore, as the PT-CT distance r reduces, the interference probability rises after falling. This is because when the CT is close to the PT, the probability in regions of interference-free is large but the probability of large delay case drops. Then, for a certain reducing step ΔG of the amplitude gain, the reduced value of the SNR becomes large. Then, when the CT obtains the test statistic Ω, it is more likely that $\gamma_{c_1} > \gamma_{c_0}$ occurs, which increases the interference probability. In the following simulations, $K = 0.8$ is used so that the interference probability at $P_I = 0.05$ can be achieved in most situations.

Figure 4.7 shows the performance of the proposed detection method to distinguish Regions I and III, where the PR is uniformly distributed in Regions I and III. From the figure, as the threshold G_t grows from 20 dB to 80 dB, the correct detection probability P_d first increases and then decreases. Since the maximum value of P_d can be obtain at $G_t \approx 52$ dB for both curves, the threshold at $G_t = 52$ dB is used for the rest of the simulations.

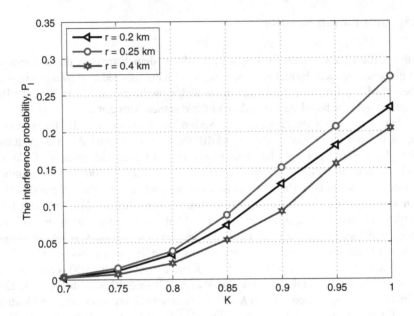

Fig. 4.6 The interference probability versus the coefficient K

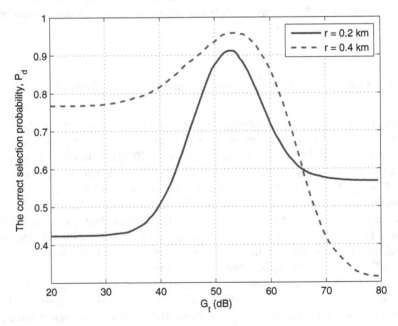

Fig. 4.7 The correct selection probability of the proposed detection method in distinguishing Regions I and III

4.5.2 Determine the Amplitude Gains

In this subsection, the amplitude gains of the two estimators are selected for Regions I and III, respectively. Here, both the successful estimation probability and the estimation error are used to evaluate the performance because the uncertainties of the wireless channel and noise may lead to the failure of the estimation, i.e., the estimators may output negative values. Thus, η is defined as the successful estimation probability, i.e.,

$$\eta = \frac{N_c}{N_s}, \tag{4.35}$$

where N_c is the number of the successful estimation values and N_s is the number of Monte Carlo trails. For the successful estimation values, ε is further defined as the estimation error, i.e.,

$$\varepsilon = \frac{|10\lg(\hat{g}_2) - 10\lg(g_2)|}{|10\lg(g_2)|}. \tag{4.36}$$

Figure 4.8 provides the performance of the estimator for the CT in Region I, where the PT-CT distance is 0.2 km. From the figure, as the amplitude gain G grows from 40 dB to 75 dB, the successful estimation probability first increases and then decreases while the estimation error first decreases and then increases. Therefore, the amplitude gain at $G = 50\,\text{dB}$ is used to obtain the minimum estimation error at $\epsilon \approx 0.03$. Meanwhile, it also achieves the large successful probability at $\eta \approx 0.9$.

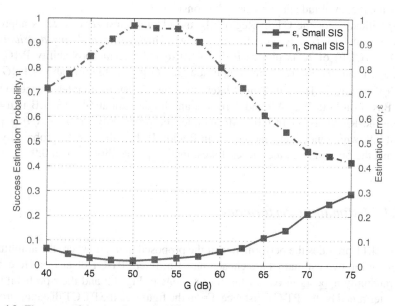

Fig. 4.8 The successful estimation probability and estimation error for the CT in Region I

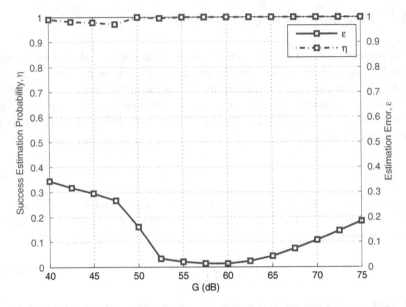

Fig. 4.9 The successful estimation probability and estimation error for the CT in Region III

Figure 4.9 provides the performance of the estimator for the CT in Region III, where the PT-CT distance is 0.2 km. From the figure, the similar trend as in Fig. 4.8 can be observed, and the minimum estimation error can be obtained at $G = 55\,\text{dB}$. However, since the estimator in Region III requires $G > \sqrt{b/g_2}$, the inequality in choosing the amplitude gain needs to be considered.

Specifically, for the CT in Region III, the obtained test statistic Ω is approximately equal to $\sqrt{b/g_2}$, i.e., $\Omega \approx \sqrt{b/g_2}$. Then, the *cumulative distribution function* (CDF) of Ω is plotted in Fig. 4.10 to indicate the probability $\Pr\{G > \sqrt{b/g_2}\}$. From the figure, as the amplitude gain G grows, the probability $\Pr\{G > \sqrt{b/g_2}\}$ increases. Then, it is more likely that the CT causes no interference to the PR. Based on both Figs. 4.9 and 4.10, the amplitude gain at $G = 65$ dB is used, where the RMSE is about 0.05. Meanwhile, the probability $\Pr\{G > \sqrt{b/g_2}\}$ is about 0.9, which indicates that the CT in Region III has about 90 % probability to conduct the probing, i.e., the successful estimation probability is about 0.9.

4.5.3   Estimation Performance

Figure 4.11 provides the performance of the proposed method, where the amplitude gains for the two estimators are chosen under different PT-CT distances. Here, the CT randomly appears in one of the four regions in Fig. 4.2 and their probabilities are determined by the PT-CT distance. From the figure, as the PT-CT distance grows

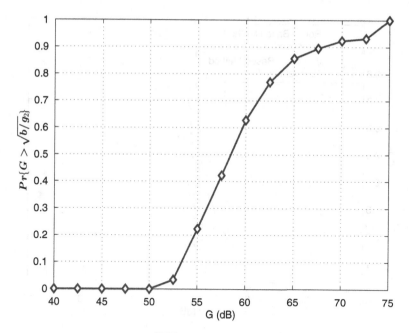

Fig. 4.10   The probability $\Pr\{G > \sqrt{b/g_2}\}$ versus the amplitude gain G

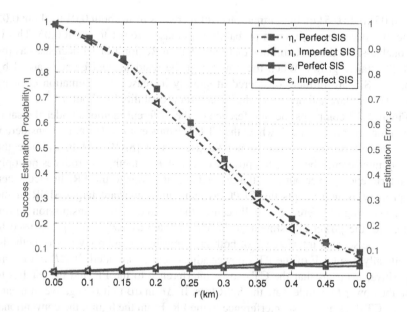

Fig. 4.11   The performance of the proposed estimation method

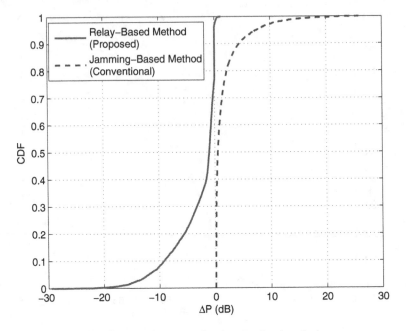

Fig. 4.12 The comparison between the proposed and conventional methods

from 0.05 km to 0.5 km, the estimation error increases from about 0.01 to about 0.05 while the successful estimation probability decreases from 1 to about 0.05. This is reasonable since the probabilities that the CT is in Region I or III decreases as the PT-CT distance grows. When the impacts of the residual interference caused by imperfect SIS are further considered, it slightly increases the estimation error by about $0.01 \sim 0.02$ and decreases the successful probability by about 0.1.

Figure 4.12 compares the interference of the different methods under the same estimation error at $\epsilon = 0.04$, where the PT-CT distance at $r = 0.2$ km is considered. Here, the interference power cannot be used as in [8, 10, 11] to evaluate the sensing interference because the proposed relay-based method introduces not only the interference power but also the desired signal power to the PR. In fact, when the primary system adopts the CLPC to maintain a constant target SINR of the PR, the sensing interference actually causes the PT to raise the transmission power. Therefore, the power adjustment at the PT, denoted as $\Delta P = p_1/p_0$, is used to evaluate the sensing interference of both jamming-based and relay-based methods. Specifically, the CDF of the PT's power adjustment is provided. If ΔP in dB unit is positive, it means that the CT causes interference to the PR and the PT has to raise the power to compensate the SNR loss. If ΔP in dB unit is negative, it means that the CT does not cause interference to the PR. From the figure, the conventional jamming-based method in [11] has 100 % probability to interfere with the PR while the proposed relay-based method has only about 5 % probability. In other words, the proposed method can reduce the interference probability by about 95 %.

It has been found that the jamming-based method obtains only 10 % interference probability in [11], which is quite smaller than the above result in 100 %. In fact, this is reasonable since the proposed method and the method in [11] use different metrics to evaluate the sensing interference. The method in [11] uses the interference power at the PR while the proposed method uses the power adjustment at the PT. In addition, the interference power threshold in [11] is relative high and thus the PR can tolerant some interference unless the interference power is above a certain value p_{peak}. In contrast, the power adjustment threshold in the proposed method is set to 0 dB. This threshold is very low and therefore the PR cannot tolerant any interference, i.e., it does not allow the CT's probing to reduce the SINR of the PR or increase the PT transmission power. Therefore, the proposed method has more strict interference definition than that in the reference [11], which leads to different interference probabilities.

Furthermore, our results in Fig. 4.12 indicate the applicable scenarios for different methods. Specifically, the proposed relay-based method is for the scenario where the primary user is very sensitive to the sensing interference. In contrast, the jamming-based method is for the scenario where the primary user is able to tolerant some sensing interference.

4.6 Summary

In this chapter, a relay-based probing method was proposed to conduct the proactive estimation. It found that depending on the location of the CT, the whole coverage of the PT can be divided into four regions and the CT that is located in two of the four regions can conduct the relay-based probing without causing interference to the PR. Thus, a detection method was developed to identify the CT located region and two estimators are designed for the two interference-free regions, respectively. Simulation results indicated that under the same estimation error, the proposed method can reduce the interference probability from 100 % to about 5 %, compared with the conventional jamming-based method.

References

1. Suraweera, H., Smith, P., & Shafi, M. (2010). Capacity limits and performance analysis of cognitive radio with imperfect channel knowledge. *IEEE Transactions on Vehicular Technology, 59*(4), 1811–1822.
2. Sboui, L., Rezki, Z., & Alouini, M. (2013). A unified framework for the ergodic capacity of spectrum sharing cognitive radio systems. *IEEE Transactions on Wireless Communications, 12*(2), 877–887.
3. Wang, Z. & Zhang, W. (2015). *Opportunistic spectrum sharing in cognitive radio networks*. New York: Springer.

4. Tannious, R. & Nosratinia, A. (2010). Cognitive radio protocols based on exploiting hybrid ARQ retransmissions. *IEEE Transactions on Wireless Communications, 9*(9), 2833–2841.
5. Hijazi, H. & Ros, L. (2009). Analytical analysis of bayesian Cramer-Rao bound for dynamical Rayleigh channel complex gains estimation in OFDM system. *IEEE Transactions on Signal Processing, 57*(5), 1889–1900.
6. Hijazi, H. & Ros, L. (2010). Joint data QR-detection and Kalman estimation for OFDM time-varying Rayleigh channel complex gains. *IEEE Transactions on Communications, 58*(1), 170–178.
7. Lioumpas, A. & Karagiannidis, G. (2010). Variable-rate M-PSK communications without channel amplitude estimation. *IEEE Transactions on Communications, 58*(5), 1477–1484.
8. Zhang, R. & Liang, Y. (2008). Exploiting hidden power-feedback loops for cognitive radio. In *Proceedings of IEEE International Symposium on New Frontiers in Dynamic Spectrum Access Networks (DySPAN 2008)*, Chicago, IL, pp. 1–5, 14–17 October 2008.
9. Zhao, G., Li, Y., & Yang, C. (2019). Proactive detection of spectrum holes in cognitive radio. In *Proceedings of IEEE International Conference on Communications (ICC 2009)*, Dresden, Germany, pp. 1–5, 14–18 June 2009.
10. Zhang, R. (2010). On active learning and supervised transmission of spectrum sharing based cognitive radios by exploiting hidden primary radio feedback. *IEEE Transactions on Communications, 58*(10), 2960–2970.
11. Bajaj, I. & Gong, Y. (2011). Cross-channel estimation using supervised probing and sensing in cognitive radio networks. In *Proceedings of IEEE International Communications Conference (ICC 2011)*, Kyoto, Japan, pp. 1–5, 5–9 June 2011.
12. Noam, Y. & Goldsmith, A. (2013). Blind null-space learning for MIMO underlay cognitive radio with primary user interference adaptation. *IEEE Transactions on Wireless Communications, 12*(4), 1722–1734.
13. Noam, Y., Manolakos, A., & Goldsmith, A. (2014). Null space learning with interference feedback for spatial division multiple access. *IEEE Transactions on Wireless Communications, 13*(10), 5699–5715.
14. Yuan, F., Villardi, G., Kojima, F., & Yano, K. (2015). Channel direction information probing for multi-antenna cognitive radio system. In *Proceedings of IEICE Society Conference*, Akashi, Japan, pp. 39–44, May 2015.
15. 3GPP TR 25.814. (2006). *hysical layer aspects for evolved universal terrestrial radio access (UTRA)*, 2006.
16. Zuleita, K. & Eduard, J. (2012). Instantaneous relaying: optimal strategies and interference neutralization. *IEEE Transactions on Signal Processing, 60*(12), 6655–6668.
17. Kang, Y., & Cho, J. (2009). Capacity of MIMO wireless channel with full-duplex amplify-and-forward relay. In *Proceeding of IEEE International Symposium on Personal, Indoor and Mobile Radio Communications (PIMRC 2009)*, Tokyo, Japan, pp. 117–121, 13–16 September 2009.
18. Riihonen, T., Werner, S., & Wichman, R. (2011). Transmit power optimization for multi-antenna decode-and-forward relays with loopback self-interference from full-duplex operation. In *Proceedings of Asilomar Conference*, Pacific Grove, CA, pp. 1408–1412, 6–9 November 2011.
19. Chen, Y. & Rapajic, P. (2010) Decentralized wireless relay network channel modeling: an analogous approach to mobile radio channel characterization. *IEEE Transactions on Communications, 58*(2), 467–473.
20. Zhao, G., Huang, B., Li, L., & Zhou, X. (2016). Relay-assisted cross-channel gain estimation for spectrum sharing. *IEEE Transactions on Communications, 64*(3), 973–986.
21. Riihonen, T., Werner, S., & Wichman, R. (2011). Mitigation of loopback self-interference in full-duplex MIMO relays. *IEEE Transactions on Signal Processing, 59*(12), 5983–5993.
22. Jain, M. et al. (2011). Practical, real-time, full duplex wireless. In *Proceedings of the 17th Annual International Conference on Mobile Radio Communications (MobiCom 2011)*, Las Vegas, Nevada, USA, pp. 1–12, 19–23 September 2011.

Chapter 5
Conclusions

Cognitive radio is regarded as one of the most promising techniques for future wireless communication systems since it can effectively enhance the spectrum utilization efficiency. To enable the coexistence of the primary and cognitive users, spectrum sensing needs to acquire the information related to the primary receiver. However, the existing spectrum sensing techniques are not applicable to address this challenge since they are designed to obtain the information related to primary transmitter.

In this Brief, three advanced spectrum sensing techniques are presented for cognitive radio. First, we addressed the problem of receiver detection in cognitive femtocell networks. In order to enable the cognitive femtocells to access the busy frequency bands of the macrocells, each femtocell needs to detect its nearby active macro users. Two detectors were proposed by exploiting the hidden information from macrocell signals, which can provide more spectrum opportunities for femtocells. Secondly, we considered the probing technique to enhance the performance of the receiver detection since the probing is able to artificially trigger the link adaption between primary transceivers. Different probing signals and detection algorithms were developed for static and dynamic scenarios, respectively. Finally, we further considered the full-duplex relay technique, which can not only improve the sensing performance but also reduce the potential interference caused by the probing signal. In particular, if a cognitive user is capable of conducting the full-duplex relay, it can estimate the cross-channel information between cognitive transmitter and primary receiver. Then, the performance of the cognitive radio system can be significantly improved by enabling the coexistence between primary and cognitive users.

Next, we discuss some other future directions of the advanced sensing techniques. To achieve high area spectrum efficiency, the tiered heterogenous networks attract much attention in recent years. The applications are not only for future cellular networks but also for future industrial networks with a very large number of devices or sensors. The current interference management in tiered heterogenous

© The Author(s) 2017
G. Zhao et al., *Advanced Sensing Techniques for Cognitive Radio*,
SpringerBriefs in Electrical and Computer Engineering,
DOI 10.1007/978-3-319-42784-3_5

networks is mainly based on the statistic information of the network and its performance is limited by the density of the wireless devices. To further enhance the performance, the interference map can be constructed using advanced sensing techniques. Since it provides the real-time interference information, the performance of the networks is expected to be significantly improved.

Another direction of the advanced sensing techniques is the non-cooperative receiver localization [1–4], which provides the location information of the receiver in other networks. This can be used in many commercial and public safety applications, e.g., counter-terrorism, law enforcement, patient monitoring, location-based services, data mining, etc. The main technique challenge is to exploit the hidden information of the signal from transmitters, which requires the state-of-the-art techniques, e.g., full-duplex relay, big data analytics, etc.

References

1. Chang, B., Guo, Z., Zhao, G., Chen, Z., & Li, L. (2015). Positioning receiver using full-duplex amplify-and-forward relay. In *Proceedings of IEEE Global Telecommunication Conference (GLOBECOM 2015)*, San Diego, CA, USA, pp. 1–6, 6–10 December 2015.
2. Zhao, G., Chang, B., Chen, Z., Li, L., & Liang, J. (2016). Third-party receiver positioning in wireless sensor networks. In *Proceedings of IEEE International Conference on Computer Communications Workshop (INFOCOM Workshop 2016)*, San Francisco, CA, USA, pp. 1–6, 10–15 April 2016.
3. Chang, B., Zhao, G., Chen, Z., & Li, L. (2016). Positioning third-party receiver via TDOA estimation in frequency division duplex systems. In *Proceedings of IEEE International Conference on Communications (ICC 2016)*, Kuala Lumpur, Malaysia, pp. 1–6, 23–27 May 2016.
4. Zhao, G., Tong, X., Li, L., & Zhou, X. (2016). Positioning third-party receiver by exploiting the close-loop power control in wireless networks. *IEEE Wireless Communications Letters, 5*(3), 268–271.

Printed in the United States
By Bookmasters

Printed in the United States
By Bookmasters